Lecture Notes in Mathematics

Edited by A. Dold and B. Eckmann

417

H. H. Keller

T0222291

Differential Calculus in Locally Convex Spaces

Springer-Verlag
Berlin · Heidelberg · New York 1974

Prof. Dr. Hans Heinrich Keller
Mathematisches Institut
Universität Zürich
Freiestr. 36
CH-8032 Zürich

Library of Congress Cataloging in Publication Data

Keller, Hans Heinrich, 1922-
Differential calculus in locally convex spaces.

(Lecture notes in mathematics ; 417)
Bibliography: p.
Includes index.
1. Locally convex spaces. 2. Calculus, Differential. I. Title. II. Series: Lecture notes in mathematics (Berlin) ; 417.
QA3.L28 no. 417 [QA322] 510'.8s [515'.73] 74-20715

AMS Subject Classifications (1970): 46-02, 46A05, 46G05, 58-02, 58C20

ISBN 3-540-06962-3 Springer-Verlag Berlin · Heidelberg · New York
ISBN 0-387-06962-3 Springer-Verlag New York · Heidelberg · Berlin

Offsetdruck: Julius Beltz, Hemsbach/Bergstr.

To Silvia

CONTENTS

2. FUNCTIONS OF CLASS C^p

APPENDIX: FUNCTIONS OF A REAL VARIABLE

ABSTRACT

Differential calculus for functions between locally convex spaces is based here on various concepts of continuous differentiability (resp. differentiability of class C^p), depending on preassigned topologies or convergence structures on the occurring spaces of continuous linear (resp. multilinear) mappings. Implications between these notions are discussed in the general case as well as in special cases. Relations to pre-existing differential theories are established.

INTRODUCTION

1. It is a well-known fact that the Fréchet differential calculus in normed spaces does not have a (unique) canonical extension to general topological vector spaces. Assume that E and F are Hausdorff locally convex spaces over the field $\mathbb{R}$ of the reals and that X is an open set in E - this will usually be the situation in the sequel - then practically all existing differential theories for functions

$$(1) \qquad f : X \to F$$

have been initiated with a definition of the differentiability of f at a point x of its domain X, usually based on a representation of the form

$$(2) \qquad f(x+h) = f(x) + Df(x)h + Rf(x,h)$$

for f in some neighbourhood of x. Here $Df(x)$, the derivative of f at x, is assumed to be an element of

$\mathcal{L}(E,F)$, the vector space of all continuous linear mappings from E into F ; a notion of differentiability of f at x is merely a requirement on the behaviour of the remainder $h \longmapsto Rf_x(h) := Rf(x,h)$ of f at x .

Several of the remainder conditions which have been established, though mutually inequivalent in the general case, are of the "Fréchet-type", i.e. they reduce to the Fréchet condition

$$(3) \qquad \lim_{h\to 0\in E} \|h\|^{-1}\cdot Rf_x(h) = 0 ,$$

if E is a normed space. This is notably the case with the conditions denoted here by (KE) , (KE') , (HL) , (HL') , (FB) , (G_b) , the last one being at the same time of the "Gâteaux-type" (see below).

On the other hand, the classical condition

$$(4) \qquad \lim_{t\to 0\in \mathbb{R}} t^{-1}Rf_x(th) = 0 \quad \text{for each } h \in E ,$$

yielding differentiability of f at x in the sense of Gâteaux-Lévy, is meaningful without any modifications even in arbitrary convergence vector spaces E and F . A whole spectrum of conditions has been derived from it by several authors by the additional requirement that the limit in (4) exists uniformly with respect to h in each set belonging to some given covering $\mathfrak{S}$ of E , consisting of bounded sets. The resulting remainder condition of "Gâteaux-type" will be denoted by $(G_{\mathfrak{S}})$.

A notion of differentiability of f at a point x which is intermediate between the "Fréchet-type" and the "Gâteaux-type" conditions, meaningful in arbitrary convergence vector spaces, has been introduced by Andrée Bastiani (cf. [11]): she requires that, for each $h_0 \in E$, the expression $t^{-1}Rf_x(th)$ has limit $0 \in F$ if (t,h) converges to $(0,h_0) \in \mathbb{R} \times F$. This condition is denoted by (MB).

An excellent survey on the various definitions of the derivative of a function f at a fixed point x has been given by V.I. Averbukh and O.G. Smolyanov in 1968 (cf. [6]). In 1970 the same authors have written an Appendix (cf. [8]) to the Russian translation of A. Frölicher and W. Bucher's Lecture Notes (cf. [31]); this Appendix [8] in a certain sense completes their survey article [6]. We have adopted here the notation of [8] for the remainder conditions except for those of "Gâteaux-type", for which we prefer the symbol $(G_{\mathfrak{S}})$.

2. Relatively little progress has been made so far in the theory of higher order derivatives of functions between non-normable spaces. This is not surprising since serious difficulties arise with this problem (cf. [38]). It seems indeed that a useful theory of functions of class C^p is not possible within the frame of topological vector spaces. Consider for example a function $f : X \to F$ which is differentiable at each point x of its domain X in the sense

of any one of the definitions. Its derivative then is a function

(5) $$Df : X \to \mathcal{L}(E,F) .$$

As already A. Frölicher and W. Bucher have pointed out in the Introduction to [31] a notion of continuous differentiability - or of a second order derivative - of f thus a priori depends on the choice of two things, namely 1° a remainder condition, 2° a topology or at least some convergence structure on $\mathcal{L}(E,F)$.

In analogy to the statement concerning Fréchet differentiability made at the beginning of this Introduction, and obviously closely related to it, one has the fact that the usual norm-topology on $\mathcal{L}(E,F)$ in the case of normed spaces E,F does not have a canonical extension to the wider category of locally convex spaces. What is worse: if E is not normable there does not even exist a compatible topology on the vector space $\mathcal{L}(E,F)$ such that for example the evaluation map

(6) $$ev : \mathcal{L}(E,F)\times E \to F$$

would be continuous. Therefore non topological convergence structures have been used by G. Marinescu [47], A. Bastiani [11], A. Frölicher and W. Bucher [31] and E. Binz [12] in their differential theories.

At first sight it might now seem that already for $p = 1$ there is quite a large and confusing variety of different notions of "differentiability of class C^p" for functions between locally convex spaces, the definition depending on two parameters: a remainder condition and a convergence structure. This is fortunately not the case due to a noteworthy fact which, though noticed and mentioned by some authors (cf. [47] p.173, [11] p.43 and [40])has probably not been paid attention to as much as it deserves. It is a classical result that if a function $f : X \to F$ between normed spaces has a Gâteaux-Lévy derivative $Df : X \to \mathcal{L}(E,F)$ which is continuous with respect to the norm-topology on $\mathcal{L}(E,F)$, then f is Fréchet differentiable. This assertion generalizes to arbitrary locally convex spaces in the sense - to be precised in Section 1.2. - that a Gâteaux-Lévy differentiable function f whose derivative Df is continuous with respect to a prescribed convergence structure on $\mathcal{L}(E,F)$ in fact satisfies a differentiability condition which is usually stronger than that of Gâteaux-Lévy one had started with. Most - but not all - of the remainder conditions mentioned before are obtained in this way.

An analogous statement holds in the more general case of functions of class C^p , here involving continuity of the derivative

(7) $$D^pf : X \to \mathcal{L}^p(E,F)$$

of f of order p with respect to some convergence structure on the space $\mathcal{L}^p(E,F)$ of all continuous p-linear mappings from E^p into F and, on the other hand, a property of the remainder $R_p f_x$ of order p in the Taylor's expansion of f at x (see Section 2.8.).

3. The present theory of differential calculus in locally convex spaces is strictly based on the concept of the continuously differentiable function (resp. function of class C^p), this notion depending solely on a topology or convergence structure on the occurring spaces of linear (resp. multilinear) mappings.

In Chapter 0 which is preparatory in character we introduce the $\mathfrak{S}$-topologies and various non-topological convergence structures on $\mathcal{L}^p(E,F)$ as well as certain broader locally convex spaces $\mathcal{H}^p_{\mathfrak{S}}(E,F)$ of p-linear mappings from E^p into F, $\mathfrak{S}$ being a covering of E consisting of bounded sets. The relations between these structures are discussed in the general case as well as in special cases, e.g. if E is metrizable, normable, a Schwartz space etc.

In Chapter 1 a function $f : X \to F$ is defined to be of class C^1_Λ, with respect to a given convergence structure Λ on $\mathcal{L}(E,F)$, if its Gâteaux-Lévy (i.e. directional) derivative $Df : X \to \mathcal{L}(E,F)$ exists and is continuous with respect to Λ. The relations among the resulting various notions are established and summarized in Table 1 on page 62. For each Λ the behaviour of the remainder of C^1_Λ-functions is derived. A chain rule is proven.

In Chapter 2 the definitions and results of Chapter 1 are extended to functions of class C^p_Λ for any natural integer p and for a convergence structure Λ on the occurring function spaces. A distinction has been made between $\mathfrak{S}$-topologies and those (non-topological) convergence structures which are finer than continuous convergence Λ_c. In the first case the higher derivatives $D^k f$, $0 \leq k \leq p$, of a C^p-function f are supposed to have their values in $\mathcal{H}^k_{\mathfrak{S}}(E,F)$ whereas in the latter case these may be in $\mathcal{L}^k(E,F)$. However, for any fixed p, the relationship among the various concepts of differentiability of class C^p_Λ is exactly the same as for $p = 1$. For functions of class C^p_Λ an estimate of the remainder of order p in their Taylor's expansion is derived. Finally a p-th order chain rule is given.

Differentiability of class C^∞_Λ with respect to a given convergence structure Λ on the occurring spaces of multilinear mappings then is defined in an obvious way. It turns out that even for quite arbitrary locally convex spaces E and F several of these notions coincide, thus yielding a scheme of implications (see Table 2 on page 109) which is considerably simpler than that for C^p on Table 1. From

Table 2 one concludes that in the following two cases all our concepts C^∞_Λ are identical: 1° if E is a Banach space and F an arbitrary locally convex space, 2° if E is a Fréchet space and F a normable locally convex space. Hence, in these important special cases, there is, within the frame of our theory, one unique notion of differentiability of class C^∞ for functions $f : X \to F$, X open in E, namely: f is of class C^∞ if there exist functions $D^p f : X \to \mathcal{L}^p(E,F)$, $D^0 f = f$, $p = 0,1,\ldots$, which are continuous for the topology Λ_s of simple (pointwise) convergence on $\mathcal{L}^p(E,F)$, and such that $D^{p+1}f$ is the Gâteaux-Lévy derivative of $D^p f$ for each p .

But even in the more general case where E is a Fréchet space while F may remain an arbitrary locally convex space, most of our concepts C^∞_Λ (all except C^∞_Θ and C^∞_Δ) coincide. The resulting notion of a C^∞-function between Fréchet spaces, identical in this case at the same time with Bastiani's, Frölicher-Bucher's and Yamamuro's C^∞ concepts, on the one hand, is based solely on directional derivatives and pointwise convergence and, on the other side, satisfies the chain rule and the strong remainder condition $(HL^{(p)})$ of all orders p .

One might hope that the approach to differential calculus developped here could in many cases simplify and facilitate its application, for example to differential geometry, especially if the spaces involved are Fréchet spaces as in [41] and [44].

4. The relationship between the present and some of the most important and comparable pre-existing differential theories, as far as functions between locally convex spaces are concerned, is as follows:

(1) Andrée Bastiani's "applications p fois différentiables" (1962, cf. [10,11]) are exactly our functions of class C_c^p .

(2) Frölicher-Bucher's "C_p-mappings" (1966, cf. [31]) coincide with our functions of class C_Π^p .

(3) In S. Yamamuro's recent Lecture Notes (1974, cf. [68]) differential calculus is based on "bounded differentiability", and his "C^p-mappings" are our functions of class C_b^p . .

ACKNOWLEDGEMENT

I am much indepted to E. Binz, H.R. Fischer and A. Frölicher for interesting discussions and valuable suggestions, to H. Jarchow and Miss T. Peter for pointing out errors and to S. Yamamuro for his encouraging remarks when I wrote my manuscript.

I thank W.L. Hill and F. Krakowski for linguistic advice.

I also thank Mrs. R. Boller, Mrs. B. Henop and Mrs. E. Minzloff for carefully typing this manuscript.

I want to express my special gratitude to Miss E. Hakios. She assisted me in every way when I wrote the first version and she had a painstaking job in correcting and completing the final typescript. Without her expert service and permanent help these Lecture Notes could hardly have been worked out.

Any mistakes in the text, however, are entirely my own responsibility.

0. SPACES OF MULTILINEAR MAPPINGS

0.0. The linear spaces $L^n(E,F)$ and $\mathcal{L}^n(E,F)$

Let E and F be Hausdorff locally convex topological vector spaces (abbr.: l.c.s.) over the field $\mathbb{R}$ of the real numbers. By Γ_E resp. Γ_F we always denote the sets of all continuous semi-norms in E resp. F, each directed by its natural order.

For every $n \in \mathbb{N} := \{0,1,\ldots\}$ we shall denote by $L^n(E,F)$ the $\mathbb{R}$-vector space (abbr.: v.s.) of all (not necessarily continuous) n-linear mappings from $E^n := E \times \ldots \times E$ (n factors) into F. As usual we define $L^0(E,F)$ to be the underlying v.s. of F, which is also denoted by F, and instead of $L^1(E,F)$ we write $L(E,F)$. We shall make frequent use of the following basic fact of multilinear algebra, which we formulate without proof.

0.0.1. Lemma. For every $(m,n) \in \mathbb{N}\times\mathbb{N}$ the mapping

$$\Theta^{m,n} : L^m(E,L^n(E,F)) \to L^{m+n}(E,F) ,$$

defined by

$$(\Theta^{m,n}u)h_1\ldots h_{m+n} := (uh_1\ldots h_m)h_{m+1}\ldots h_{m+n} ,$$

is an isomorphism of v.s., called canonical.

In the sequel the v.s. $L^m(E,L^n(E,F))$ will often tacitly be identified to $L^{m+n}(E,F)$ by means of $\Theta^{m,n}$. Likewise each subspace of $L^m(E,L^n(E,F))$ will be viewed as a subspace of $L^{m+n}(E,F)$.

Let us denote by $\mathcal{L}^n(E,F)$ the linear subspace of $L^n(E,F)$ consisting of all <u>continuous</u> n-<u>linear</u> <u>mappings</u> from E^n into F . Again, by definition, we have $\mathcal{L}^0(E,F) = F$ qua v.s., and we write $\mathcal{L}(E,F)$ instead of $\mathcal{L}^1(E,F)$.

0.1. $\mathfrak{S}$-topologies on $\mathcal{L}^n(E,F)$

Given a collection $\mathfrak{S}$ of bounded subsets of the l.c.s. E, such that $\mathfrak{S}$ covers E , we denote by $\Lambda_{\mathfrak{S}}$ the topology of $\mathfrak{S}$-convergence on each occurring linear function space $\mathcal{L}^n(E,F)$, F an l.c.s., $n \in \mathbb{N}$. (In $\mathcal{L}^0(E,F) = F$, of course, $\Lambda_{\mathfrak{S}}$ is the original topology of F.) We know that $\Lambda_{\mathfrak{S}}$ is a Hausdorff locally convex topology, defined by the semi-norms

$$u \longmapsto |u|_{\beta,S} := \sup\{|uh_1\ldots h_n|_\beta \,|\, h_i \in S, 1 \leq i \leq n\} ,$$

where $\beta \in \Gamma_F$ and $S \in \mathfrak{S}$. (Without loss of generality we can always assume that $S \in \mathfrak{S}$ and $S' \in \mathfrak{S}$ imply $S \cup S' \in \mathfrak{S}$. In this case the defining family $(|\ |_{\beta,S})_{\beta \in \Gamma_F, S \in \mathfrak{S}}$ of semi-norms is directed by its natural order.) Thus

$$\mathcal{L}^n_{\mathfrak{S}}(E,F) := (\mathcal{L}^n(E,F),\Lambda_{\mathfrak{S}})$$

is an l.c.s.

If $\mathfrak{S},\mathfrak{S}'$ are two covers of E by bounded sets such that $\mathfrak{S} \subset \mathfrak{S}'$, then clearly $\Lambda_{\mathfrak{S}}$ is coarser than $\Lambda_{\mathfrak{S}'}$. We shall mainly be concerned with the following special cases of $\mathfrak{S}$-topologies: the topologies Λ_s, Λ_k, Λ_{pk}, Λ_b of respectively simple, compact, precompact, bounded convergence, corresponding to $\mathfrak{S} = \mathfrak{S}_s$, $\mathfrak{S}_k$, $\mathfrak{S}_{pk}$, $\mathfrak{S}_b$, the collections of all respectively finite, compact, precompact, bounded subsets of E. Each of these topologies is finer than the preceding; Λ_s is the coarsest and Λ_b the finest of all $\mathfrak{S}$-topologies on $\mathcal{L}^n(E,F)$, $\mathfrak{S}$ being a cover of E by bounded sets. We shall denote the corresponding l.c.s. by $\mathcal{L}^n_s(E,F)$, $\mathcal{L}^n_k(E,F)$, $\mathcal{L}^n_{pk}(E,F)$, $\mathcal{L}^n_b(E,F)$ respectively.

Let E,F and n be as before. We shall also consider the topology Λ_s of simple convergence on the v.s. $L^n(E,F)$ of all n-linear mappings from E^n into F. The l.c.s. $\mathcal{L}^n_s(E,F)$ is then a subspace of the l.c.s.

$$L^n_s(E,F) := (L^n(E,F),\Lambda_s) .$$

As a matter of fact, Λ_s is induced by the (product-) topology on $F^{(E^n)}$, the l.c.s. of all mappings from E^n into F.

0.1.1. <u>Lemma</u>. Let E and F be l.c.s. and let $\mathfrak{S}$ denote a cover of E by bounded sets. Then for every $(m,n) \in \mathbb{N}\times\mathbb{N}$ the l.c.s. $\mathcal{L}^{m+n}_{\mathfrak{S}}(E,F)$ is a subspace of the l.c.s. $\mathcal{L}^m_{\mathfrak{S}}(E,\mathcal{L}^n_{\mathfrak{S}}(E,F))$.

Here the underlying v.s. $\mathcal{L}^m(E,\mathcal{L}^n_{\mathfrak{S}}(E,F))$ of the latter l.c.s. is identified to a linear subspace of $L^{m+n}(E,F)$ by means of $\Theta^{m,n}$.

Proof. Assume $v \in \mathcal{L}^{m+n}(E,F)$. There is a (unique) $u \in L^m(E,\mathcal{L}^n(E,F))$ such that $\Theta^{m,n}u = v$. Given $\beta \in \Gamma_F$ there exists an $\alpha \in \Gamma_E$ such that

$$|v|_{\beta,\alpha} := \sup\{|vh_1\ldots h_{m+n}|_\beta \,\big|\, |h_i|_\alpha \leq 1, 1 \leq i \leq m+n\} < \infty .$$

We obviously have

$$\sup\{|(uh_1\ldots h_m)h_{m+1}\ldots h_{m+n}|_\beta \,\big|\, |h_i|_\alpha \leq 1, 1 \leq i \leq m+n\} = |v|_{\beta,\alpha} .$$

Therefore, for every $S \in \mathfrak{S}$ and every $(h_1,\ldots,h_m) \in E^m$ we get

$$|uh_1\ldots h_m|_{\beta,S} := \sup\{|(uh_1\ldots h_m)k_1\ldots k_n|_\beta \,\big|\, k_j \in S, 1 \leq j \leq n\}$$
$$\leq \mu^n \cdot |v|_{\beta,\alpha} \cdot |h_1|_\alpha \cdot \ldots \cdot |h_m|_\alpha ,$$

where

$$\mu := \sup\{|k_j|_\alpha \,\big|\, k_j \in S , 1 \leq j \leq n\} .$$

As $\beta \in \Gamma_F$ and $S \in \mathfrak{S}$ are arbitrary, this shows that the m-linear mapping $u : E^m \to \mathcal{L}^n_{\mathfrak{S}}(E,F)$ is continuous, thus $u \in \mathcal{L}^m(E,\mathcal{L}^n_{\mathfrak{S}}(E,F))$.

From the formulae above we may also derive that for every $\beta \in \Gamma_F$ and every $S \in \mathfrak{S}$ we have $|v|_{\beta,S} = |u|_{(\beta,S),S}$ whenever $v = \Theta^{m,n}u$; here we have used the semi-norms of $v \in \mathcal{L}^{m+n}_{\mathfrak{S}}(E,F)$ respectively $u \in \mathcal{L}^m_{\mathfrak{S}}(E,\mathcal{L}^n_{\mathfrak{S}}(E,F))$, determined by β and S , namely

$$|v|_{\beta,S} := \sup\{|vh_1\ldots h_{m+n}|_\beta \,|\, h_i \in S, 1 \le i \le m+n\} ,$$

$$|u|_{(\beta,S),S} := \sup\{|uh_1\ldots h_m|_{\beta,S} \,|\, h_i \in S, 1 \le i \le m\} .$$

This means that $\mathcal{L}^{m+n}_{\mathfrak{S}}(E,F)$ carries the topology induced by $\mathcal{L}^{m}_{\mathfrak{S}}(E,\mathcal{L}^{n}_{\mathfrak{S}}(E,F))$.

0.1.2. Lemma. Let E be a metrizable l.c.s., F an arbitrary l.c.s. and $n \in \mathbb{N}$. Let X be a metrizable topological space and assume that $g : X \to \mathcal{L}^n_k(E,F)$ is a continuous function. Then the map $\tilde{g} : X \times E^n \to F$, associated to g , defined by

$$\tilde{g}(x,h_1,\ldots,h_n) := g(x)h_1\ldots h_n ,$$

is continuous.

Proof. If $\tilde{g}$ were not continuous we could choose $\beta \in \Gamma_F$, $\varepsilon > 0$, a convergent sequence $(x_j)_{j\in\mathbb{N}}$ with limit x in X and for each i , $1 \le i \le n$, a convergent sequence $(h_i^{(j)})_{j\in\mathbb{N}}$ with limit h_i in E , such that

$$|g(x_j)h_1^{(j)}\ldots h_n^{(j)} - g(x)h_1\ldots h_n|_\beta > \varepsilon$$

for every $j \in \mathbb{N}$. Now the sets $\{h_i^{(j)} | j \in \mathbb{N}\} \cup \{h_i\}$, $1 \le i \le n$, are compact. Therefore, on the other hand, according to the hypotheses of the lemma and the continuity of the n-linear map $g(x) : E^n \to F$, there would exist $p \in \mathbb{N}$, such that

$$|(g(x_j)-g(x))h_1^{(j)}\dots h_n^{(j)}|_\beta \leq \frac{\varepsilon}{2}$$

and

$$|g(x)h_1^{(j)}\dots h_n^{(j)}-g(x)h_1\dots h_n| \leq \frac{\varepsilon}{2} ,$$

for $j \geq p$, thus contradicting the assumption made above

0.1.3. Theorem. Let E be a metrizable and F an arbitrary l.c.s., and let $\mathfrak{S}$ denote a collection of bounded subsets of E which contains all compact sets. Then for each $(m,n) \in \mathbb{N}\times\mathbb{N}$ the canonical linear isomorphism $\Theta^{m,n}$ of 0.0.1. defines an isomorphism

$$\Theta^{m,n} : \mathcal{L}^m_{\mathfrak{S}}(E,\mathcal{L}^n_{\mathfrak{S}}(E,F)) \to \mathcal{L}^{m+n}_{\mathfrak{S}}(E,F)$$

of l.c.s.

Proof. As $\Lambda_{\mathfrak{S}}$ is finer than Λ_k , by 0.1.2. if $u \in \mathcal{L}^m(E,\mathcal{L}^n_{\mathfrak{S}}(E,F))$ then the $(m+n)$-linear map

$$\Theta^{m,n}u = \tilde{u} : E^{m+n} \to F$$

is continuous, thus $\Theta^{m,n}u \in \mathcal{L}^{m+n}(E,F)$. Together with 0.1.1. we get the desired result.

0.1.4. Lemma. Let E be a metrizable and barrelled l.c.s., F an arbitrary l.c.s. and $n \in \mathbb{N}$. If X is a metrizable topological space and if $g : X \to \mathcal{L}^n_s(E,F)$ is a continuous function, then the map $\tilde{g} : X\times E^n \to F$, associated to g , is continuous.

Proof. For $n = 1$ we can use the same argument as in the proof of 0.1.2., taking into account that by the Banach-Steinhaus theorem every simply convergent sequence in $\mathcal{L}(E,F)$ is convergent with respect to Λ_{pk}. For arbitrary $n \in \mathbb{N}$ we proceed by induction.

0.1.5. Theorem. Let E be a metrizable and barrelled l.c.s., F an arbitrary l.c.s. and $\mathfrak{S}$ any cover of E by bounded sets. Then for every $(m,n) \in \mathbb{N}\times\mathbb{N}$ the canonical linear isomorphism of 0.0.1. induces an isomorphism

$$\Theta^{m,n} : \mathcal{L}^m_{\mathfrak{S}}(E,\mathcal{L}^n_{\mathfrak{S}}(E,F)) \to \mathcal{L}^{m+n}_{\mathfrak{S}}(E,F)$$

of l.c.s.

Proof. As $\Lambda_{\mathfrak{S}}$ is finer than Λ_s and coarser than Λ_b, the assertion of 0.1.5. follows from 0.1.4. with the arguments used in the proof of 0.1.3.

0.2. The l.c.s. $\mathcal{H}^n_{\mathfrak{S}}(E,F)$

Let E and F be l.c.s., let $\mathfrak{S}$ be a cover of E consisting of bounded sets. The l.c.s. $\mathcal{H}^n_{\mathfrak{S}}(E,F)$, $n \in \mathbb{N}$, are defined inductively by

$$\mathcal{H}^0_{\mathfrak{S}}(E,F) := F$$

$$\mathcal{H}^n_{\mathfrak{S}}(E,F) := \mathcal{L}_{\mathfrak{S}}(E,\mathcal{H}^{n-1}_{\mathfrak{S}}(E,F)) \quad \text{if } n \geq 1 .$$

0.2.1. <u>Proposition</u>. For every $(m,n) \in \mathbb{N}\times\mathbb{N}$ we have

$$\mathcal{H}^m_{\mathfrak{S}}(E,\mathcal{H}^n_{\mathfrak{S}}(E,F)) = \mathcal{H}^{m+n}_{\mathfrak{S}}(E,F) .$$

Proof. By induction on m .

0.2.2. <u>Lemma</u>. For every $n \in \mathbb{N}$ there exists a unique injective linear map

$$\Psi^n : \mathcal{H}^n_{\mathfrak{S}}(E,F) \to L^n(E,F) ,$$

called canonical, defined recursively by

$$\text{(i) } \Psi^o := id_F$$

$$\text{(ii) } \Psi^n u := \Theta^{1,n-1}(\Psi^{n-1} \circ u) \quad \text{for every} \quad u \in \mathcal{H}^n_{\mathfrak{S}}(E,F) .$$

In the sequel each $\mathcal{H}^n_{\mathfrak{S}}(E,F)$ qua v.s. will tacitly be identified to its image by means of Ψ^n , and the elements of $\mathcal{H}^n_{\mathfrak{S}}(E,F)$ will therefore be considered as n-<u>linear mappings</u> from E^n into F , $\mathfrak{S}$-hypocontinuous in a certain sense.

Proof. Trivially, Ψ^o is a unique injective linear map. Assume that

$$\Psi^{n-1} : \mathcal{H}^{n-1}_{\mathfrak{S}}(E,F) \to L^{n-1}(E,F)$$

has been defined as an injective linear map. Given $u \in \mathcal{H}^n_{\mathfrak{S}}(E,F) = \mathcal{L}_{\mathfrak{S}}(E,\mathcal{H}^{n-1}_{\mathfrak{S}}(E,F))$ we have $\Psi^{n-1} \circ u \in L(E,L^{n-1}(E,F))$ and $\Theta^{1,n-1}(\Psi^{n-1} \circ u) \in L^n(E,F)$. Therefore Ψ^n is well-defined by (ii), and one easily verifies that Ψ^n is linear and injective.

0.2.3. Proposition. Let E and F be l.c.s. and let $\mathfrak{S}$ denote a cover of E by bounded sets. Then for every $n \in \mathbb{N}$ the l.c.s. $\mathcal{L}^n_{\mathfrak{S}}(E,F)$ is a subspace of the l.c.s. $\mathcal{H}^n_{\mathfrak{S}}(E,F)$.

Here the underlying v.s. of the latter l.c.s. is identified to a linear subspace of $L^n(E,F)$ by means of Ψ^n .

Proof. By induction on n . We have $\mathcal{L}^0_{\mathfrak{S}}(E,F) = F = \mathcal{H}^0_{\mathfrak{S}}(E,F)$. Assume the assertion of 0.2.3. to be true for $n-1$. By 0.1.1. $\mathcal{L}^n_{\mathfrak{S}}(E,F)$ is a subspace of $\mathcal{L}_{\mathfrak{S}}(E,\mathcal{L}^{n-1}_{\mathfrak{S}}(E,F))$, and by the induction hypothesis the latter is a subspace of $\mathcal{L}_{\mathfrak{S}}(E,\mathcal{H}^{n-1}_{\mathfrak{S}}(E,F)) = \mathcal{H}^n_{\mathfrak{S}}(E,F)$.

0.2.4. Proposition. Let E and F be l.c.s. and $n \in \mathbb{N}$. Assume that $\mathfrak{S}$ and $\mathfrak{S}'$ are two covers of E by bounded sets such that $\mathfrak{S} \subset \mathfrak{S}'$. Then $\mathcal{H}^n_{\mathfrak{S}'}(E,F)$, qua v.s., is a linear subspace of $\mathcal{H}^n_{\mathfrak{S}}(E,F)$, and the inclusion map is continuous.

Proof. This is trivial if $n = 0$. Assume it to be true for $n-1$. Then $\mathcal{H}^n_{\mathfrak{S}'}(E,F) = \mathcal{L}_{\mathfrak{S}'}(E,\mathcal{H}^{n-1}_{\mathfrak{S}'}(E,F))$ and $\mathcal{L}_{\mathfrak{S}}(E,\mathcal{H}^{n-1}_{\mathfrak{S}'}(E,F))$ have the same underlying v.s., but the topology of the first is finer. From the induction hypothesis it follows that $\mathcal{L}_{\mathfrak{S}}(E,\mathcal{H}^{n-1}_{\mathfrak{S}'}(E,F))$ is a linear subspace of $\mathcal{L}_{\mathfrak{S}}(E,\mathcal{H}^{n-1}_{\mathfrak{S}}(E,F)) = \mathcal{H}^n_{\mathfrak{S}}(E,F)$ and that its topology is finer than the topology induced from the latter space.

If $\mathfrak{S} = \mathfrak{S}_s, \mathfrak{S}_k, \mathfrak{S}_{pk}, \mathfrak{S}_b$ respectively we shall denote $\mathcal{H}^n_{\mathfrak{S}}(E,F)$ by $\mathcal{H}^n_s(E,F)$, $\mathcal{H}^n_k(E,F)$, $\mathcal{H}^n_{pk}(E,F)$, $\mathcal{H}^n_b(E,F)$ respectively. We then have the following

0.2.5. <u>Corollary</u>. Let E and F be l.c.s. and $n \in \mathbb{N}$. Then $\mathcal{H}^n_b(E,F) \subset \mathcal{H}^n_{pk}(E,F) \subset \mathcal{H}^n_k(E,F) \subset \mathcal{H}^n_s(E,F)$, and the inclusion maps are continuous.

The space $\mathcal{H}^n_s(E,F)$ consists just of those n-linear maps from E^n into F which are <u>separately</u> <u>continuous</u>; its topology is induced from $F^{(E^n)}$.

0.2.6. <u>Theorem</u>. Let E be a metrizable and F an arbitrary l.c.s. Let $\mathfrak{S}$ be a collection of bounded subsets of E which contains all compact sets. Then for every $n \in \mathbb{N}$ we have

$$\mathcal{H}^n_{\mathfrak{S}}(E,F) = \mathcal{L}^n_{\mathfrak{S}}(E,F) .$$

Proof. By induction on n, using 0.1.3.

0.2.7. <u>Theorem</u>. Let E be a metrizable and barrelled l.c.s., let F be an arbitrary l.c.s. For every cover $\mathfrak{S}$ of E by bounded sets and for every $n \in \mathbb{N}$ we have

$$\mathcal{H}^n_{\mathfrak{S}}(E,F) = \mathcal{L}^n_{\mathfrak{S}}(E,F) .$$

Proof. By induction on n, using 0.1.5.

0.3. The structure Λ_c of continuous convergence

Let E and F be two l.c.s. For every $n \in \mathbb{N}$ we can consider the evaluation map

$$ev : \mathcal{L}^n(E,F) \times E^n \to F ,$$

defined by

$$ev(u,(h_1,\ldots,h_n)) := uh_1\ldots h_n .$$

It is known that ev is not continuous for any v.s. topology on $\mathcal{L}^n(E,F)$ whatsoever, unless E is normable or $F = \{0\}$ or $n = 0$. Neither will, except in special cases, composition

$$\pi : \mathcal{L}(F,G) \times \mathcal{L}(E,F) \to \mathcal{L}(E,G) ,$$

defined by $\pi(v,u) := v \circ u$, for three l.c.s. E, F and G, be a continuous (bilinear) mapping, if the occurring function spaces are endowed with l.c. topologies. For these statements the reader is referred to [37] and [38].

In addition to the $\mathfrak{S}$-topologies we are therefore going to use certain convergence structures ("Limitierungen") in the sense of H.R. Fischer [30], called "pseudo-topologies" in A. Frölicher and W. Bucher's Lecture Notes [31]. If Λ is a convergence structure on $\mathcal{L}^n(E,F)$ we shall denote by $\Lambda(u)$ the set of those filters on $\mathcal{L}^n(E,F)$ which converge to $u \in \mathcal{L}^n(E,F)$ with respect to Λ. The arising convergence space ("Limesraum" in [30], "pseudo-topological space" in [31]) $(\mathcal{L}^n(E,F),\Lambda)$ will be denoted by $\mathcal{L}^n_\Lambda(E,F)$. The basic notions in the theory of convergence spaces are assumed to be known. For a detailed exposition the reader is referred to [30]. We always consider topologies as special cases of convergence structures.

We recall that a convergence structure Λ on a v.s. is said to be compatible iff both, addition and scalar multiplication, are (jointly) continuous. In this case Λ is invariant with respect to translations, thus determined by the set $\Lambda(0)$ of those filters which converge to 0. A v.s. endowed with a separated compatible convergence structure is called a convergence vector space (abbr.: c.v.s.). Any Hausdorff topological v.s., in particular any l.c.s., is a c.v.s.

The convergence structure Λ_c of continuous convergence on $\mathcal{L}^n(E,F)$ is defined to be the coarsest of those convergence structures (compatible or not) on $\mathcal{L}^n(E,F)$, which make the evaluation map ev continuous. It should be emphasized that this definition makes sense whenever E and F are two c.v.s. We shall write

$$\mathcal{L}^n_c(E,F) := (\mathcal{L}^n(E,F),\Lambda_c) .$$

In this section a short account on Λ_c will be given, as far as it will be needed in the sequel. For further information the reader is referred to [13].

Let E and F be two c.v.s. and $n \in \mathbb{N}$. By definition of Λ_c a filter $\mathcal{F}$ on $\mathcal{L}^n(E,F)$ converges to an element $u \in \mathcal{L}^n(E,F)$ with respect to Λ_c, i.e., $\mathcal{F} \in \Lambda_c(u)$, iff for every filter $\mathcal{R}$ on E^n which converges to some $h \in E^n$ the "evaluated" filter $\mathcal{F}(\mathcal{R}) := ev(\mathcal{F}\times\mathcal{R})$ converges to $u(h)$ in F. From this remark one easily derives the following universal mapping property which characterizes Λ_c :

0.3.1. Theorem. Let E and F be two c.v.s. and $n \in \mathbb{N}$. Let X be an arbitrary convergence space. A mapping $g : X \to \mathcal{L}^n_c(E,F)$ is continuous iff the associated map $\tilde{g} : X \times E^n \to F$ is continuous.

Proof. If g is continuous, then $\tilde{g} := ev \circ (g \times id_{E^n})$ is continuous because ev is continuous. Assume conversely that $\tilde{g}$ is continuous. Given filters $\mathcal{X}$ and $\mathcal{R}$ on X and E^n which converge to x and h respectively, by the assumption $g(\mathcal{X})(\mathcal{R}) = \tilde{g}(\mathcal{X} \times \mathcal{R})$ converges to $\tilde{g}(x,h) = g(x)h$. As $h \in E^n$ was arbitrary this means that $g(\mathcal{X})$ converges to $g(x)$. Thus g is continuous at x, hence continuous on X, as x was arbitrary.

0.3.2. Corollary. Given a convergence structure Λ on $\mathcal{L}^n(E,F)$, compatible or not, the following two statements are equivalent:

(i) Λ is finer than Λ_c.

(ii) For every convergence space X and every continuous mapping $g : X \to \mathcal{L}^n_\Lambda(E,F)$ the associated map $\tilde{g} : X \times E^n \to F$ is continuous.

Proof. Obviously (i) implies (ii). In order to get the converse assertion, apply (ii) to $X = \mathcal{L}^n_\Lambda(E,F)$ and g the identity, thus $\tilde{g} = ev$.

0.3.3. Proposition. Let E and F be c.v.s. For every $n \in \mathbb{N}$ the space $\mathcal{L}_c^n(E,F)$ is a c.v.s.

Proof. $\mathcal{L}^n(E,F)$ is a linear subspace of the v.s. of all continuous mappings from E^n into F, on which the structure of continuous convergence is compatible and separated (cf. [13]).

0.3.4. Theorem. Let E and F be two c.v.s. For every $(m,n) \in \mathbb{N} \times \mathbb{N}$ the canonical linear isomorphism

$$\Theta^{m,n} : L^m(E,L^n(E,F)) \to L^{m+n}(E,F)$$

induces an isomorphism

$$\Theta^{m,n} : \mathcal{L}_c^m(E,\mathcal{L}_c^n(E,F)) \to \mathcal{L}_c^{m+n}(E,F)$$

of c.v.s.

Proof. From the universal mapping property (0.3.1.) of Λ_c it follows that $\Theta^{m,n}$ maps the v.s. $\mathcal{L}^m(E,\mathcal{L}_c^n(E,F))$ isomorphically onto the v.s. $\mathcal{L}^{m+n}(E,F)$. In order to see that this is even an isomorphism of c.v.s. we apply the criterion of continuity (0.3.1.) to $\Theta^{m,n}$ and to its inverse. The details are left to the reader.

0.3.5. Theorem. Let E, F and G be three c.v.s. Assume $n \in \mathbb{N}$ and let $m_1,\ldots,m_n$ be elements of $\mathbb{N}$ with sum m. Then the canonical multilinear mapping

$$\pi : \mathcal{L}_c^n(F,G) \times \prod_{1\leq i\leq n} \mathcal{L}_c^{m_i}(E,F) \to \mathcal{L}_c^m(E,G) ,$$

defined by

$$\pi(v,(u_1,\ldots,u_n)) := v \circ (u_1\times\ldots\times u_n) ,$$

is continuous.

Proof. According to 0.3.1. we have to prove that the map

$$\tilde{\pi} : \mathcal{L}_c^n(F,G) \times \prod_{1\le i\le n} \mathcal{L}_c^{m_i}(E,F) \times E^m \to G ,$$

associated to π , is continuous. Now, as

$$\prod_{1\le i\le n} \mathcal{L}_c^{m_i}(E,F) \times E^m = \prod_{1\le i\le n} (\mathcal{L}_c^{m_i}(E,F) \times E^{m_i}) ,$$

$\tilde{\pi}$ can be factorized as follows

$$\mathcal{L}_c^n(F,G) \times \prod_{1\le i\le n} (\mathcal{L}_c^{m_i}(E,F) \times E^{m_i}) \to \mathcal{L}_c^n(F,G)\times F^n \to G,$$

where the arrows are received by executing evaluations, which are continuous.

In the remainder of this section we shall be concerned with the special case that E and F are l.c.s. We might emphasize that nevertheless Λ_c is not a topology on $\mathcal{L}^n(E,F)$ unless E has finite dimension or $F = \{0\}$ or $n = 0$. Let $\mathfrak{U}$ denote the neighbourhood filter of 0 in E, and let Γ_E, Γ_F denote the sets of all continuous semi-norms in E, F respectively.

0.3.6. Proposition. Let E be a topological v.s., F a c.v.s. and $n \in \mathbb{N}$. A filter $\mathcal{F}$ on $\mathcal{L}^n(E,F)$ is a member of $\Lambda_c(0)$ iff the following two conditions are satisfied:

(1) $\mathcal{F} \in \Lambda_s(0)$.

(2) $\mathcal{F}(\mathfrak{U}^n)$ converges to 0 in F .

Proof. (a) Assume that $\mathcal{F} \in \Lambda_c(0)$. Given $h \in E^n$, the filters $\mathcal{F}([h])$ and $\mathcal{F}(\mathfrak{U}^n)$ both converge to 0 in F . (Here $[h]$ denotes the trivial ultrafilter on E^n generated by $\{h\}$.)

(b) Assume conversely that (1) and (2) are fulfilled. In case $n = 1$, if $h \in E$ then $\mathcal{F}(\mathcal{U}+[h])$ is finer than $\mathcal{F}(\mathcal{U}) + \mathcal{F}([h])$ and therefore converges to 0 in F. As $h \in E$ is arbitrary this means $\mathcal{F} \in \Lambda_c(0)$.

In order to carry out the proof for $n > 1$ it suffices to look at the case $n = 2$. Assume that $h = (h_1,h_2) \in E \times E$. We have to prove that $\mathcal{F}((\mathcal{U}+[h_1]) \times (\mathcal{U}+[h_2]))$ is convergent to 0 in F. It is easy to see that this last filter on F is finer than

$$\mathcal{F}(\mathcal{U}\times\mathcal{U}) + \mathcal{F}(\mathcal{U}\times[h_2]) + \mathcal{F}([h_1] \times \mathcal{U}) + \mathcal{F}([h_1] \times [h_2]) .$$

Now $\mathcal{F}([h_1] \times [h_2]) = \mathcal{F}([h])$ and $\mathcal{F}(\mathcal{U}\times\mathcal{U})$ are convergent to 0 in F by the hypothesis. In order to see that the same statement holds for $\mathcal{F}(\mathcal{U}\times[h_2])$ we take into account that $\mathbb{V}\cdot\mathcal{U} = \mathcal{U}$, where $\mathbb{V}$ denotes the filter of 0-neighbourhoods on $\mathbb{R}$. We then have

$$\mathcal{F}(\mathcal{U}\times[h_2]) = \mathcal{F}(\mathbb{V}\cdot\mathcal{U}\times[h_2]) = \mathcal{F}(\mathcal{U}\times\mathbb{V}\cdot[h_2]) .$$

Therefore $\mathcal{F}(\mathcal{U}\times[h_2])$, and likewise $\mathcal{F}([h_1]\times\mathcal{U})$, converges to 0 in F .

We are going to reformulate 0.3.6. in terms of semi-norms in case E and F are l.c.s. For this purpose we recall the following notation already introduced in the proof of 0.1.1. If $u \in L^n(E,F)$, $\alpha \in \Gamma_E$ and $\beta \in \Gamma_F$ we shall write

$$|u|_{\beta,\alpha} := \sup\{|uh_1\ldots h_n|_\beta \,\big|\, |h_i|_\alpha \leq 1, 1 \leq i \leq n\} \leq \infty .$$

We know that $u \in \mathcal{L}^n(E,F)$ iff for every $\beta \in \Gamma_F$ there exists an $\alpha \in \Gamma_E$, such that $|u|_{\beta,\alpha} < \infty$.

0.3.7. Corollary. Let E and F be l.c.s. and $n \in \mathbb{N}$. For a filter $\mathcal{F}$ on $\mathcal{L}^n(E,F)$ we have $\mathcal{F} \in \Lambda_c(0)$ iff the following two conditions are satisfied:

(1) $\mathcal{F} \in \Lambda_s(0)$.

(2') $(\forall \beta \in \Gamma_F)(\exists \alpha \in \Gamma_E)(\exists Q \in \mathcal{F}) \qquad \sup_{u \in Q} |u|_{\beta,\alpha} < \infty$.

Proof. We have to prove that the conditions (2) in 0.3.6. and (2') in 0.3.7. are equivalent.

(a) Assume that $\mathcal{F}(\mathcal{U}^n)$ converges to 0 in F. Given $\beta \in \Gamma_F$ there exist $\alpha \in \Gamma_E$, $\delta > 0$ and $Q \in \mathcal{F}$ such that $h_i \in E$, $|h_i|_\alpha \leq \delta$, $1 \leq i \leq n$, and $u \in Q$ imply $|uh_1 \dots h_n|_\beta \leq 1$. From this we conclude

$$\sup_{u \in Q} |u|_{\beta,\alpha} \leq \delta^{-n} < \infty .$$

(b) If we assume that (2') above is fulfilled, then, given $\beta \in \Gamma_F$ and $\varepsilon > 0$, we choose $\alpha \in \Gamma_E$ and $Q \in \mathcal{F}$ such that

$$\mu := \sup_{u \in Q} |u|_{\beta,\alpha} < \infty .$$

Now for every $u \in \mathcal{L}^n(E,F)$ and every $(h_1,\dots,h_n) \in E^n$ such that $u \in Q$ and $|h_i|_\alpha \leq (\varepsilon\mu^{-1})^{\frac{1}{n}}$, $1 \leq i \leq n$, we have

$$|uh_1 \dots h_n|_\beta \leq |u|_{\beta,\alpha} \cdot |h_1|_\alpha \cdot \dots \cdot |h_n|_\alpha \leq \varepsilon .$$

As β and c were arbitrary this means that $\mathcal{F}(\mathcal{U}^n)$ converges to 0 in F.

We are now going to compare Λ_c to $\mathfrak{S}$-topologies on $\mathcal{L}^n(E,F)$. The following results will be important in the sequel.

0.3.8. Theorem. Let E and F be l.c.s. On each v.s. $\mathcal{L}^n(E,F)$, $n \in \mathbb{N}^{\cdot}$, the structure Λ_c of continuous convergence is finer than the topology Λ_{pk} of precompact convergence.

Proof. Assume $\mathcal{F} \in \Lambda_c(0)$. According to condition (2') of 0.3.7., given $\beta \in \Gamma_F$ there exist $\alpha \in \Gamma_E$ and $Q \in \mathcal{F}$ such that $\mu := \sup_{u \in Q} |u|_{\beta,\alpha} < \infty$. Let P be a precompact set in E. Define ρ to be the larger of the real numbers 1 and $\sup_{h \in P} |h|_\alpha$.

For every δ, $0 < \delta \leq 1$, we choose the finite set $M_\delta \subset P$ such that

$$P \subset M_\delta + \{h \in E \mid |h|_\alpha \leq \delta\} .$$

Given $h_i \in P$, $1 \leq i \leq n$, we can therefore find points $h_i' \in M_\delta$, $1 \leq i \leq n$, such that $|h_i - h_i'|_\alpha \leq \delta$, and we also have $|h_i'|_\alpha \leq \rho$, $1 \leq i \leq n$, because $h_i' \in P$. If $u \in Q$, a short computation yields the inequality

$$|uh_1 \ldots h_n - uh_1' \ldots h_n'|_\beta \leq 2^n \cdot \rho^{n-1} \cdot \delta \cdot \mu .$$

Given $\varepsilon > 0$ arbitrary, if we choose $\delta \leq 2^{-n-1} \cdot \rho^{1-n} \cdot \mu^{-1} \cdot \varepsilon$, the expression above has value $\leq \frac{1}{2}\varepsilon$. By condition (1) of 0.3.7. there is $Q_1 \in \mathcal{F}$ such that

$$\sup\{|uh_1'\ldots h_n'|_\beta \mid u\in Q_1, h_i'\in M_\delta, 1\leq i\leq n\} \leq \frac{1}{2}\varepsilon .$$

Then $u \in Q \cap Q_1 \in \mathcal{F}$ and $h_i \in P$, $1 \leq i \leq n$, imply $|uh_1\ldots h_n|_\beta \leq \varepsilon$. As $\beta \in \Gamma_F$, $\varepsilon > 0$ and the precompact set $P \subset E$ were arbitrary, this proves that $\mathcal{F}$ converges to $0 \in \mathcal{L}^n(E,F)$ with respect to Λ_{pk}.

0.3.9. Theorem. Let E and F be l.c.s. and $n \in \mathbb{N}$. Assume that E is metrizable (resp. metrizable and barrelled). Let X be a metrizable topological space and let $g : X \to \mathcal{L}^n_k(E,F)$ (resp. $g : X \to \mathcal{L}^n_s(E,F)$) be a continuous mapping. Then $g : X \to \mathcal{L}^n_c(E,F)$ is continuous.

Proof. This is an immediate consequence of 0.1.2. (resp. 0.1.4.), if we take into account the universal mapping property 0.3.1. of Λ_c.

0.4. The structure Λ_{qb} of quasi-bounded convergence

Let E be a c.v.s.

A set $B \subset E$ is called bounded if the filter $\mathbb{V}\cdot B$ converges to $0 \in E$. (Here, as always, $\mathbb{V}$ denotes the filter of 0-neighbourhoods in $\mathbb{R}$.) If E is an l.c.s., this concept of boundedness coincides with the usual one. A filter $\mathcal{B}$ on E will be called a bounded filter, if it contains a bounded set. Clearly, if B is a bounded set, then $[B]$, the filter on E generated by $\{B\}$, is a bounded filter.

A. Frölicher and W. Bucher (cf. [31], 2.5.) define a filter $\mathcal{B}$ on E to be a quasi-bounded filter iff $\mathbb{V}\cdot\mathcal{B}$ converges to $0 \in E$. Each bounded and each convergent filter on E is quasi-bounded. On the other hand one has

0.4.1. Proposition. For a c.v.s. E the following two statements are equivalent:

(i) Each quasi-bounded filter on E is bounded.

(ii) Each convergent filter on E is bounded.

Proof. (i) implies (ii) because each convergent filter is quasi-bounded. Assume now that E has property (ii) and that $\mathcal{B}$ is a quasi-bounded filter on E. As $\mathbb{V}\cdot\mathcal{B}$ converges to $0 \in E$, by (ii) there exist $\delta > 0$ and $B \in \mathcal{B}$ such that $[-\delta, +\delta]\cdot B$ is bounded. But then B is a bounded set and $\mathcal{B}$ a bounded filter.

For convenience a c.v.s. E will be said to have property (QB) if (i), and thus (ii), is true. An l.c.s. has property (QB) iff it is normable. A general c.v.s. may have property (QB) without being a normable l.c.s.

Let E be a c.v.s. and $n \in \mathbb{N}$. It is easy to see that in the c.v.s. E^n a set is bounded iff it is included in some set of the form B^n where B is bounded in E. A filter on E^n is bounded (resp. quasi-bounded) iff it is finer than some filter of the form $\mathcal{B}^n$, where $\mathcal{B}$ is a bounded (resp. quasi-bounded) filter on E. From this we conclude that E^n has property (QB) iff E has.

Let E and F be two c.v.s. and $n \in \mathbb{N}$. We define the structures Λ_b (resp. Λ_{qb}) of bounded convergence (resp. quasi-bounded convergence) on the v.s. $\mathcal{L}^n(E,F)$ by the following requirements:

For every filter $\mathcal{F}$ on $\mathcal{L}^n(E,F)$ and every $u \in \mathcal{L}^n(E,F)$ $\mathcal{F} \in \Lambda_b(u)$, (resp. $\mathcal{F} \in \Lambda_{qb}(u)$)

iff $(\mathcal{F}-u)(\mathcal{B}^n)$ converges to $0 \in F$ for every bounded (resp. quasi-bounded) filter $\mathcal{B}$ on E.

We leave it to the reader to verify that Λ_b and Λ_{qb}, defined as above, are separated convergence structures on the v.s. $\mathcal{L}^n(E,F)$, which are by definition invariant with respect to translations. We shall write

$$\mathcal{L}^n_b(E,F) := (\mathcal{L}^n(E,F),\Lambda_b), \quad \mathcal{L}^n_{qb}(E,F) := (\mathcal{L}^n(E,F),\Lambda_{qb})$$

0.4.2. Proposition. Let E and F be c.v.s. For every $n \in \mathbb{N}$ the spaces $\mathcal{L}^n_b(E,F)$ and $\mathcal{L}^n_{qb}(E,F)$ are c.v.s.

Proof. The verification of the compatibility conditions (cf. [31], §2, §6) for $\Lambda_b(0)$ and $\Lambda_{qb}(0)$ is straightforward.

0.4.3. Proposition. Let E be a c.v.s. and F an l.c.s. Then for every $n \in \mathbb{N}$ the space $\mathcal{L}^n_b(E,F)$ is an l.c.s.

Proof. The semi-norms (cf. Sect.0.1.)

$$u \longmapsto |u|_{\beta,B} := \sup\{|uh_1 \ldots h_n|_\beta \,|\, h_i \in B, 1 \leq i \leq n\}$$

on $\mathcal{L}^n(E,F)$, where $\beta \in \Gamma_F$ and B is a bounded set in E , define a separated l.c. topology $\bar{\Lambda}_b$ on $\mathcal{L}^n(E,F)$. By definition of Λ_b , for a filter $\mathcal{F}$ on $\mathcal{L}^n(E,F)$ one has $\mathcal{F} \in \Lambda_b(0)$ iff $\mathcal{F}$ converges to 0 with respect to $\bar{\Lambda}_b$

Remark. If both, E and F , are l.c.s. then Λ_b coincides with the <u>topology of bounded convergence</u> on $\mathcal{L}^n(E,F)$, introduced in Sect.0.1. This justifies the use of the same symbol "Λ_b" .

The structure Λ_{qb} of quasi-bounded convergence has first been used by Frölicher and Bucher (cf. [31], §6). For general information about Λ_{qb} we refer the reader to [31]. Here we shall only summarize those properties of Λ_{qb} which will be used in the sequel, especially concerning the cases where E or F , or both, are l.c.s.

0.4.4. <u>Theorem</u>. Let E and F be c.v.s. For every $n \in \mathbb{N}$ the convergence structure Λ_{qb} on $\mathcal{L}^n(E,F)$ is finer than Λ_b . If E has property (QB) (e.g. if E is a normable l.c.s.) then $\Lambda_b = \Lambda_{qb}$.

Proof. Trivial.

0.4.5. Lemma. Let E be a metrizable l.c.s., F an arbitrary l.c.s. and $n \in \mathbb{N}$. Assume $\mathcal{F}$ is a convergent filter on $\mathcal{L}_b^n(E,F)$ with a countable basis. Then $\mathcal{F}$ still converges in $\mathcal{L}_{qb}^n(E,F)$ to the same point.

Proof. For the sake of simplicity we take $n = 1$. The general case is treated in the same way.

It suffices to consider a filter $\mathcal{F} \in \Lambda_b(0)$ which possesses a countable basis $(Q_i)_{i \in \mathbb{N}}$ such that $Q_{i+1} \subset Q_i$ for every $i \in \mathbb{N}$. If we assume $\mathcal{F} \notin \Lambda_{qb}(0)$, then there exist a neighbourhood V of 0 in F and a quasi-bounded filter $\mathcal{B}$ on E such that $Q_i(B) \not\subset V$ for every $i \in \mathbb{N}$ and every $B \in \mathcal{B}$.

Let $(U_j)_{j \in \mathbb{N}}$ be a neighbourhood basis of 0 in E, consisting of absolutely convex sets, and such that $U_{j+1} \subset U_j$ for every $j \in \mathbb{N}$. By the definition of quasi-boundedness, for each $j \in \mathbb{N}$, there is a $B_j \in \mathcal{B}$ which is absorbed by U_j, and we can assume $B_{j+1} \subset B_j$ for each $j \in \mathbb{N}$. From the hypotheses on V and $\mathcal{B}$ it follows that $Q_i(B_j) \not\subset V$ for every pair $(i,j) \in \mathbb{N} \times \mathbb{N}$. We can therefore choose $h_j \in B_j$ such that $Q_j(h_j) \not\subset V$ for every $j \in \mathbb{N}$. Consider the set $B := \{h_j | j \in \mathbb{N}\}$. It is bounded in E because, given any $p \in \mathbb{N}$, the finite set $\{h_0, \ldots, h_{p-1}\}$ as well as the set $\{h_j | j \geq p\} \subset B_p$ are absorbed by U_p.

From the construction of B it follows that $Q_i(B) \not\subset V$ for every $i \in \mathbb{N}$. This contradicts the hypothesis that $\mathcal{F} \in \Lambda_b(0)$.

0.4.6. Corollary. Let E, F and n be as in 0.4.5. Every convergent sequence in $\mathcal{L}^n_b(E,F)$ converges in $\mathcal{L}^n_{qb}(E)$ to the same point.

0.4.7. Theorem. Let E be a metrizable l.c.s., F an arbitrary l.c.s. and $n \in \mathbb{N}$. Assume that X is a metrizable topological space and that $g : X \to \mathcal{L}^n_b(E,F)$ is a continuous mapping. Then $g : X \to \mathcal{L}^n_{qb}(E,F)$ is continuous.

Proof. For every $x \in X$ the neighbourhood filter $\mathcal{X}$ of x and hence its image $g(\mathcal{X})$ in $\mathcal{L}^n(E,F)$ has a countable basis.

0.4.8. Theorem. Let E and F be c.v.s. For every $n \in \mathbb{N}$ the structure Λ_{qb} of quasi-bounded convergence on $\mathcal{L}^n(E,F)$ is finer than the structure Λ_c of continuous convergence.

Proof. Assume $\mathcal{F} \in \Lambda_{qb}(0)$. Let $\mathcal{R}$ be any convergent filter on E^n. Then $\mathcal{R}$ is quasi-bounded, hence $\mathcal{F}(\mathcal{R})$ converges to 0 in F. This means $\mathcal{F} \in \Lambda_c(0)$.

0.4.9. Corollary. For every $n \in \mathbb{N}$ the canonical bilinear mapping

$$ev : \mathcal{L}^n_{qb}(E,F)\times E^n \to F$$

is continuous.

0.4.10. <u>Corollary</u>. Let E and F be l.c.s., $F \neq \{0\}$. For every $n \in \mathbb{N}$, $n \neq 0$, the convergence structure Λ_{qb} on $\mathcal{L}^n(E,F)$ is a topology iff E is normable.

Proof. If E is normable then, by 0.4.3. and 0.4.4., $\Lambda_{qb} = \Lambda_b$ is a topology. If E is not normable, there does not exist a compatible topology on $\mathcal{L}^n(E,F)$ which is finer than Λ_c (cf. [37], 4.2.).

We close this section with two statements on canonical mappings between spaces of continuous multilinear mappings endowed with quasi-bounded convergence.

0.4.11. <u>Theorem</u>. Let E be an l.c.s. and F a c.v.s. For every $(m,n) \in \mathbb{N}\times\mathbb{N}$ the canonical linear isomorphism

$$\Theta^{m,n} : L^m(E,L^n(E,F)) \to L^{m+n}(E,F)$$

induces an isomorphism

$$\Theta^{m,n} : \mathcal{L}^m_{qb}(E,\mathcal{L}^n_{qb}(E,F)) \to \mathcal{L}^{m+n}_{qb}(E,F)$$

of c.v.s.

Proof. (1) Since Λ_{qb} is finer than Λ_c , $\Theta^{m,n}$ maps the v.s. $\mathcal{L}^m(E,\mathcal{L}^n_{qb}(E,F))$ isomorphically onto a linear subspace of the v.s. $\mathcal{L}^{m+n}(E,F)$.

(2) Let $v \in \mathcal{L}^{m+n}(E,F)$ be arbitrary. There exists exactly one $u \in L^m(E,L^n(E,F))$ such that $\Theta^{m,n}u = v$. Clearly

$u \in L^m(E,\mathcal{L}^n(E,F))$. We claim that the m-linear map $u : E^m \to \mathcal{L}^n_{qb}(E,F)$ is continuous. As E is an l.c.s., continuity at $0 \in E^m$ will do. We therefore have to show that $u(\mathcal{U}^m)$ converges to 0 in $\mathcal{L}^n_{qb}(E,F)$ where $\mathcal{U}$ is the neighbourhood filter of 0 in E , or that $u(\mathcal{U}^m)(\mathcal{B}^n)$ converges to 0 in F for any quasi-bounded filter $\mathcal{B}$ on E . This is indeed true because

$$u(\mathcal{U}^m)(\mathcal{B}^n) = v(\mathcal{U}^m \times \mathcal{B}^n) = v(\mathbb{V}^n \cdot \mathcal{U}^m \times \mathcal{B}^n) = v(\mathcal{U}^m \times (\mathbb{V} \cdot \mathcal{B})^n) .$$

Thus $u \in \mathcal{L}^m(E,\mathcal{L}^n_{qb}(E,F))$.

(3) Now we know that $\Theta^{m,n}$ induces an isomorphism of v.s. from $\mathcal{L}^m(E,\mathcal{L}^n_{qb}(E,F))$ onto $\mathcal{L}^{m+n}(E,F)$. In order to show that this is an isomorphism of c.v.s., with respect to Λ_{qb} on both spaces, we have only to observe that for any filter $\mathcal{F}$ on $\mathcal{L}^m(E,\mathcal{L}^n_{qb}(E,F))$ and any filter $\mathcal{B}$ on E we have

$$(\mathcal{F}(\mathcal{B}^m))(\mathcal{B}^n) = (\Theta^{m,n}(\mathcal{F}))(\mathcal{B}^{m+n}) .$$

From this last equation we conclude that $\mathcal{F}$ converges to 0 (with respect to Λ_{qb}) iff $\Theta^{m,n}(\mathcal{F})$ does. This proves the assertion.

0.4.12. <u>Theorem</u>. Let E , F, G be three l.c.s. and $n \in \mathbb{N}$. Let $m_1,\ldots,m_n$ be elements of $\mathbb{N}$ with sum m . The canonical $(n+1)$-linear mapping

$$\pi : \mathcal{L}^n_{qb}(F,G) \times \prod_{1\leq i\leq n} \mathcal{L}^{m_i}_{qb}(E,F) \to \mathcal{L}^m_{qb}(E,G) ,$$

defined by

$$\pi(v,(u_1,\dots,u_n)) := v\circ(u_1\times\dots\times u_n) ,$$

is continuous.

Proof. For the sake of simplicity we take $n = 1$. The general case can be treated with the same method. We now have to prove that composition as a bilinear mapping

$$\pi : \mathcal{L}_{qb}(F,G)\times\mathcal{L}^m_{qb}(E,F) \to \mathcal{L}^m_{qb}(E,G)$$

is continuous.

Let us assume that $\mathcal{F}$ and $\mathcal{G}$ are convergent filters on $\mathcal{L}^m_{qb}(E,F)$ and $\mathcal{L}_{qb}(F,G)$ respectively with limits u , v say. We have to show that the filter $\mathcal{G}\circ\mathcal{F} := \pi(\mathcal{G}\times\mathcal{F})$ on $\mathcal{L}^m_{qb}(E,G)$ converges to $v\circ u$. One immediately verifies that the filter $\mathcal{G}\circ\mathcal{F} - v\circ u$ is finer than

$$(\mathcal{G}-v)\circ u + v\circ(\mathcal{F}-u) + (\mathcal{G}-v)\circ(\mathcal{F}-u) .$$

We claim that each term of this sum is convergent to 0 in $\mathcal{L}^m_{qb}(E,G)$. In order to see this, assume that $\mathcal{B}$ is any quasi-bounded filter on E , hence $\mathcal{B}^m$ a quasi-bounded filter on E^m . It follows that $(\mathcal{F}-u)(\mathcal{B}^m)$ is convergent to 0 in F and that $u(\mathcal{B}^m)$ is quasi-bounded in F because

$$\mathcal{V}\cdot u(\mathcal{B}^m) = \mathcal{V}^m\cdot u(\mathcal{B}^m) = u((\mathcal{V}\cdot\mathcal{B})^m)$$

converges to $0 \in F$. Therefore, if we apply each term of the sum above to $\mathcal{B}^m$ we receive three filters on G , each of which converges to $0 \in G$. This proves everything.

0.5. The convergence structure Π on $\mathcal{L}^n(E,F)$

0.5.1. Proposition. Let E and F be two l.c.s. For every $n \in \mathbb{N}$ there exists a unique separated compatible convergence structure Π on $\mathcal{L}^n(E,F)$ such that the set $\Pi(0)$ of the filters $\mathcal{F}$ on $\mathcal{L}^n(E,F)$ which converge to 0 is characterized as follows:

$$\mathcal{F} \in \Pi(0) \quad \text{iff} \quad (\forall\beta\epsilon\Gamma_F)(\exists\alpha\epsilon\Gamma_E)(\forall\varepsilon>0)(\exists Q\epsilon\mathcal{F}) \\ \sup\{|u|_{\beta,\alpha} \mid u \in Q\} \leq \varepsilon .$$

Proof. One has to verify (1) that $\Pi(0)$ is the set of filters convergent to 0 for some convergence structure on $\mathcal{L}^n(E,F)$; (2) that $\Pi(0)$ satisfies the compatibility conditions (cf. [31], 2.4.2.) and (3) that Π is separated. The proofs are straightforward.

We shall use the following notation:

$$\mathcal{L}^n_\Pi(E,F) := (\mathcal{L}^n(E,F),\Pi) .$$

0.5.2. Theorem. Let E and F be l.c.s. For every $n \in \mathbb{N}$ the convergence structure Π on $\mathcal{L}^n(E,F)$ is finer than the structure Λ_{qb} of quasi-bounded convergence.

Proof. Suppose $\mathcal{F} \in \Pi(0)$ and let $\mathcal{B}$ be a quasi-bounded filter on E . We have to show that $\mathcal{F}(\mathcal{B})$ converges to 0 in F . Given $\beta \in \Gamma_F$ we choose $\alpha \in \Gamma_E$ according to the definition of $\Pi(0)$ in 0.5.1. By the definition of quasi-boundedness there is a $B \in \mathfrak{B}$ and a $\mu > 0$ such that $|h|_\alpha \leq \mu$ for every $h \in B$. If now $\varepsilon > 0$ is given, we can choose $Q \in \mathcal{F}$ such that $|u|_{\beta,\alpha} \leq \varepsilon \cdot \mu^{-n}$ for every $u \in Q$. Hence $u \in Q$ and $(h_1,\ldots,h_n) \in B^n$ imply

$$|uh_1\ldots h_n|_\beta \leq |u|_{\beta,\alpha} \cdot |h_1|_\alpha \cdot \ldots \cdot |h_n|_\alpha \leq \varepsilon .$$

As $\beta \in \Gamma_F$ and $\varepsilon > 0$ were arbitrary, this proves everything.

0.5.3. <u>Corollary</u>. Π is finer than every $\mathfrak{S}$-topology ($\mathfrak{S}$ a collection of bounded sets on E) and finer than the structure Λ_c of continuous convergence on $\mathcal{L}^n(E,F)$.

0.5.4. <u>Corollary</u>. The canonical (n+1)-linear mapping

$$ev : \mathcal{L}^n_\Pi(E,F) \times E^n \to F$$

is continuous.

0.5.5. <u>Proposition</u>. Let E be a normable and F an arbitrary l.c.s. On each v.s. $\mathcal{L}^n(E,F)$, $n \in \mathbb{N}$, we have

$$\Pi = \Lambda_{qb} = \Lambda_b .$$

Proof. We know that $\Lambda_{qb} = \Lambda_b$ (0.4.4.). Let $h \longmapsto \|h\|$ denote an admissable norm on E . Assume $\mathcal{F} \in \Lambda_b(0) = \Lambda_{qb}(0)$. As $B := \{h \in E \mid \|h\| \leq 1\}$ is bounded, $\mathcal{F}(B^n)$ converges to 0 in F . Thus for every $\beta \in \Gamma_F$ and $\varepsilon > 0$ there is a $Q \in \mathcal{F}$ such that $\sup\{|uh_1\ldots h_n|_\beta \mid \|h_i\| \leq 1, 1 \leq i \leq n\} \leq \varepsilon$ for each $u \in Q$. This means $\mathcal{F} \in \Pi(0)$.

0.5.6. Corollary. Let E and F be l.c.s., $F \neq \{0\}$ and $n \in \mathbb{N}$, $n \geq 1$. The convergence structure Π on $\mathcal{L}^n(E,F)$ is a topology iff E is normable.

Proof. If E is normable, then Π is the topology Λ_b of bounded convergence. If E is not normable, then Π, being compatible and finer than Λ_c, is not a topology (cf. [37], 4.2.).

0.5.7. Theorem. Let E and F be l.c.s. If E is a Schwartz space, then for every $n \in \mathbb{N}$ we have

$$\Pi = \Lambda_{qb} = \Lambda_c$$

on $\mathcal{L}^n(E,F)$.

Proof. Assume that $\mathfrak{F} \in \Lambda_c(0)$. Due to 0.3.7. (2'), given $\beta \in \Gamma_F$ there are an $\alpha_0 \in \Gamma_E$ and a $Q_0 \in \mathfrak{F}$ such that $|u|_{\beta,\alpha_0}$ is bounded on Q_0. Since E is a Schwartz space we can find an $\alpha \in \Gamma_E$ with the property that for every $\delta > 0$ there exists a finite set $M_\delta \subset E$ such that

$$U_\alpha := \{h \in E \mid |h|_\alpha \leq 1\} \subset M_\delta + \{h \in E \mid |h|_{\alpha_0} \leq \delta\} ,$$

i.e. the 0-neighbourhood U_α is totally bounded (precompact if α_0 is a norm) in the v.s. E endowed with the l.c. topology generated by the single semi-norm α_0. We now can use the same arguments as in the proof of 0.3.8. in order to show that for every $\varepsilon > 0$ there is a $Q \in \mathfrak{F}$ such that

$u \in Q$ and $h_i \in U_\alpha$, $1 \leq i \leq n$, imply $|uh_1 \ldots h_n|_\beta \leq \varepsilon$, i.e. $u \in Q$ implies $|u|_{\beta,\alpha} \leq \varepsilon$. Here $\beta \in \Gamma_F$ and $\varepsilon > 0$ are quite arbitrary, while $\alpha \in \Gamma_E$ only depends on β. This means $\mathcal{F} \in \Pi(0)$.

According to the definition of A. Frölicher and W. Bucher (cf. [31], 2.5., 2.6.) a filter $\mathcal{F}$ on a v.s. is said to be <u>equable</u> iff $\mathcal{F} = \mathbb{V}\cdot\mathcal{F}$. Here, as always, $\mathbb{V}$ denotes the filter of all 0-neighbourhoods on $\mathbb{R}$. A c.v.s. and its convergence structure Λ are called equable, iff for every $\mathcal{F} \in \Lambda(0)$ there exists an equable filter $\mathcal{G} \in \Lambda(0)$ which is coarser than $\mathcal{F}$. Every Hausdorff topological v.s. is an equable c.v.s. Any multilinear mapping from a finite product of equable c.v.s. into a c.v.s. is continuous iff it is continuous at 0.

For every separated compatible convergence structure Λ on a v.s. there exists among all equable compatible convergence structures on this v.s. which are finer than Λ a coarsest one; it is called the <u>equable convergence structure associated to</u> Λ (and denoted by Λ^*). One has

$$\Lambda^*(0) = \{\mathcal{F} \in \Lambda(0) \mid \mathcal{F} = \mathbb{V}\cdot\mathcal{F}\} .$$

For a detailed exposition we refer to Frölicher and Bucher's Lecture Notes [31] in whose theory of higher derivatives equable c.v.s. play a major rôle.

0.5.8. <u>Proposition</u>. Let E and F be l.c.s. For every $n \in \mathbb{N}$ the c.v.s. $\mathcal{L}^n_\Pi(E,F)$ is equable.

Proof. Assume $\mathcal{F} \in \Pi(0)$. We claim that there exists a filter $\mathcal{A} = \mathbb{V}\cdot\mathcal{A} \in \Pi(0)$ which is coarser than $\mathcal{F}$.

According to the definition of Π there is a function $\varphi : \Gamma_F \to \Gamma_E$ with the property that for every $\beta \in \Gamma_F$ and every $\varepsilon > 0$ there exists a $Q \in \mathcal{F}$ such that $\sup_{u \in Q} |u|_{\beta,\varphi(\beta)} \leq \varepsilon$. For each $\beta \in \Gamma_F$ we put

$$A_\beta := \{u \in \mathcal{L}^n(E,F) \mid |u|_{\beta,\varphi(\beta)} \leq 1\} .$$

Let $\mathcal{A}$ be the filter on $\mathcal{L}^n(E,F)$ which is generated by the sets $\varepsilon \cdot A_\beta$, $\beta \in \Gamma_F, \varepsilon > 0$. Then, as is easily seen, $\mathcal{A}$ has the required properties.

0.5.9. Lemma. Let E and F be l.c.s. and $n \in \mathbb{N}$. Let $\mathcal{F}$ be a filter on $\mathcal{L}^n(E,F)$ such that $\mathbb{V} \cdot \mathcal{F} \notin \Pi(0)$. Then $\mathcal{F} \notin \Lambda_c(0)$.

Proof. From the hypothesis on $\mathcal{F}$ one immediately concludes

$$(\exists \beta \in \Gamma_F)(\forall \alpha \in \Gamma_E)(\forall Q \in \mathcal{F}) \quad \sup_{u \in Q} |u|_{\beta,\alpha} = +\infty .$$

This is exactly the negation of the condition (2') in corollary (0.3.7.).

0.5.10. Corollary. If the filter $\mathcal{F}$ is convergent in $\mathcal{L}^n_c(E,F)$ then $\mathcal{F}$ is quasi-bounded in $\mathcal{L}^n_\Pi(E,F)$, hence also quasi-bounded in $\mathcal{L}^n_{qb}(E,F)$.

Proof. Assume $\mathcal{F} \in \Lambda_c(u)$ for some $u \in \mathcal{L}^n(E,F)$, thus $\mathcal{F}-u \in \Lambda_c(0)$. From 0.5.9. we conclude that $\mathbb{V} \cdot (\mathcal{F}-u) \in \Pi(0)$. The assertion of the corollary follows from the fact that $\mathbb{V} \cdot \mathcal{F}$ is finer than $\mathbb{V} \cdot (\mathcal{F}-u) + \mathbb{V} \cdot u$ and that $\mathbb{V} \cdot u \in \Pi(0)$.

0.5.11. Proposition. Let E and F be l.c.s. and $n \in \mathbb{N}$. If $\mathcal{F}$ is an equable filter on $\mathcal{L}^n(E,F)$ and $\mathcal{F} \in \Lambda_c(0)$, then $\mathcal{F} \in \Pi(0)$.

Proof. This is an immediate consequence of 0.5.9.

0.5.12. Theorem. Let E and F be l.c.s. and $n \in \mathbb{N}$. Then Π is the equable convergence structure associated to Λ_c and to Λ_{qb} on $\mathcal{L}^n(E,F)$, i.e. $\Lambda_c^{\#} = \Lambda_{qb}^{\#} = \Pi$.

Proof. As $\Pi(0) \subset \Lambda_{qb}(0) \subset \Lambda_c(0)$, from 0.5.11. it follows that we have

$$\Pi(0) = \{\mathcal{F} \in \Lambda_c(0) \,|\, \mathcal{F} = \mathbb{V}\cdot\mathcal{F}\} = \{\mathcal{F} \in \Lambda_{qb}(0) \,|\, \mathcal{F} = \mathbb{V}\cdot\mathcal{F}\} .$$

0.5.13. Theorem. Let E, F, G be three l.c.s. and $n \in \mathbb{N}$. Let $m_1,\ldots,m_n$ be elements of $\mathbb{N}$ with sum m. Then the canonical $(n+1)$-linear mapping

$$\pi : \mathcal{L}^n_\Pi(F,G) \times \prod_{1 \le i \le n} \mathcal{L}^{m_i}_\Pi(E,F) \to \mathcal{L}^m_\Pi(E,G) ,$$

defined by $\pi(v,(u_1,\ldots,u_n)) := v \circ (u_1 \times \ldots \times u_n)$, is continuous.

Proof. Since Π is equable it suffices to prove that the multilinear map π is continuous at 0. This follows from the fact that for continuous semi-norms α, β, γ in E, F, G resp., $u_i \in \mathcal{L}^{m_i}(E,F)$, $1 \le i \le n$, and $v \in \mathcal{L}^n(F,G)$ we have

$$|\pi(v,(u_1,\ldots,u_n))|_{\gamma,\alpha} \le |v|_{\gamma,\beta}\cdot|u_1|_{\beta,\alpha}\cdot\ldots\cdot|u_n|_{\beta,\alpha} .$$

The details are left to the reader.

0.5.14. Proposition. Let E , F and G be l.c.s. . For every covering $\mathfrak{S}$ of E consisting of bounded sets the composition map

$$\pi : \mathcal{L}_{\Pi}(F,G) \times \mathcal{L}_{\mathfrak{S}}(E,F) \to \mathcal{L}_{\mathfrak{S}}(E,G)$$

is continuous.

Proof. Since Π and $\Lambda_{\mathfrak{S}}$ are equable again it suffices to prove that the bilinear map π is continuous at the origin. The verification is now based on the inequality

$$|v \circ u|_{\gamma,S} \leq |v|_{\gamma,\beta} \cdot |u|_{\beta,S}$$

which holds for arbitrary $u \in \mathcal{L}(E,F)$, $v \in \mathcal{L}(F,G)$, $\beta \in \Gamma_F$, $\gamma \in \Gamma_G$ and $S \in \mathfrak{S}$.

0.6. Marinescu's convergence structure Δ

In [47] G. Marinescu endows $\mathcal{L}^n(E,F)$, E and F being l.c.s., with a structure of a "réunion pseudotopologique d'espaces localement convexes", which turns out to be a compatible convergence structure. In this section we shall mention a few results on this structure denoted by Δ , although it will not play a major rôle in the differential calculus to be developed in the forthcoming chapters. For further information the reader is referred to [37], where Δ is denoted by Λ .

Let E and F denote two l.c.s. and $n \in \mathbb{N}$. In order to define Marinescu's convergence structure Δ on $\mathcal{L}^n(E,F)$ we introduce the set Φ of all mappings from Γ_F into Γ_E; we shall endow Φ with the partial ordering inherited from Γ_E. (For $\alpha,\alpha' \in \Gamma_E$ we have $\alpha \leq \alpha'$ if $|h|_\alpha \leq |h|_{\alpha'}$, for every $h \in E$; for $\phi,\phi' \in \Phi$, by definition, we have $\phi \leq \phi'$ iff $\phi(\beta) \leq \phi'(\beta)$ for all $\beta \in \Gamma_F$.) Then, as Γ_E is directed by $\leq$, so is Φ.

For every $\phi \in \Phi$ we consider the linear subspace

$$\mathcal{L}^n_\phi(E,F) := \{u \in \mathcal{L}^n(E,F) \,|\, |u|_{\beta,\phi(\beta)} < \infty \text{ for every } \beta \in \Gamma_F\}$$

of the v.s. $\mathcal{L}^n(E,F)$, endowed with the l.c. topology determined by the semi-norms $u \longmapsto |u|_{\beta,\phi(\beta)}$, $\beta \in \Gamma_F$. It is easy to see that $(\mathcal{L}^n_\phi(E,F))_{\phi\in\Phi}$ is an inductive system of l.c.s.: if $\phi \leq \phi'$, then $\mathcal{L}^n_\phi(E,F) \subset \mathcal{L}^n_{\phi'}(E,F)$, and the inclusion map is linear and continuous. We also have

$$\mathcal{L}^n(E,F) = \bigcup_{\phi\in\Phi} \mathcal{L}^n_\phi(E,F) \qquad \text{qua v.s.}$$

Now Marinescu's structure Δ is defined to be the finest of all those convergence structures on $\mathcal{L}^n(E,F)$ for which every inclusion map

$$\mathcal{L}^n_\phi(E,F) \to \mathcal{L}^n(E,F) \qquad (\phi \in \Phi)$$

is continuous. This means that the convergence space

$$\mathcal{L}^n_\Delta(E,F) := (\mathcal{L}^n(E,F),\Delta)$$

is the inductive limit of the inductive system $(\mathcal{L}^n_\phi(E,F))_{\phi\in\Phi}$

in the category of convergence spaces (not in the category of topological spaces or in the category of l.c.s., except in special cases).

0.6.1. Proposition. Let E and F be l.c.s. For every $n \in \mathbb{N}$ the convergence space $\mathcal{L}^n_\Delta(E,F)$ is a c.v.s.

A filter $\mathcal{F}$ on $\mathcal{L}^n(E,F)$ is a member of $\Delta(0)$, the set of filters converging to 0 with respect to Δ , iff there exists a $\phi \in \Phi$ such that the following conditions are satisfied:

(1) $\mathcal{L}^n_\phi(E,F) \in \mathcal{F}$.

(2) The trace $\mathcal{F}_\phi$ of $\mathcal{F}$ on $\mathcal{L}^n_\phi(E,F)$ converges to 0 with respect to the topology of $\mathcal{L}^n_\phi(E,F)$.

Remark. From (1) it follows that $\mathcal{F}_\phi$ is a basis of $\mathcal{F}$ in $\mathcal{L}^n(E,F)$.

Proof. One verifies that $\Delta(0)$, as characterized by the second part of 0.6.1., defines a separated compatible convergence structure Δ on the v.s. $\mathcal{L}^n(E,F)$ such that $\Delta(0)$ is exactly the set of filters which converge to 0 with respect to Δ . (Such verifications have repeatedly been carried out; we therefore leave the details to the reader.) It then follows that for every filter $\mathcal{F}$ on $\mathcal{L}^n(E,F)$ and for every $u \in \mathcal{L}^n(E,F)$ we have $\mathcal{F} \in \Delta(u)$ iff there exists a $\phi \in \Phi$ such that $u \in \mathcal{L}^n_\phi(E,F)$ and $\mathcal{F}$ has a basis on $\mathcal{L}^n_\phi(E,F)$ which converges to u in the topology of this l.c.s.

From this it is immediate that Δ is the finest of those convergence structures on $\mathcal{L}^n(E,F)$ for which every inclusion map $\mathcal{L}^n_\phi(E,F) \to \mathcal{L}^n(E,F)$ is continuous.

0.6.2. Theorem. Let E and F be l.c.s. On each v.s. $\mathcal{L}^n(E,F)$, $n \in \mathbb{N}$, Marinescu's convergence structure Δ is finer than Π. If E or F is normable then equality holds: $\Delta = \Pi$.

Proof. An equivalent version of the second part of 0.6.1. is as follows: $\mathcal{F} \in \Delta(0)$ iff

(1) $$(\exists \phi_0 \in \Phi)(\exists Q_0 \in \mathcal{F})(\forall u \in Q_0)(\forall \beta \in \Gamma_F) \qquad |u|_{\beta,\phi_0(\beta)} < \infty .$$

(2) $$(\exists \phi \in \Phi)(\forall \beta \in \Gamma_F)(\forall \varepsilon > 0)(\exists Q \in \mathcal{F})(\forall u \in Q) \qquad |u|_{\beta,\phi(\beta)} \leq \varepsilon .$$

On the other hand we know that $\mathcal{F} \in \Pi(0)$ iff (2) holds. Therefore $\Delta(0) \subset \Pi(0)$, hence Δ is finer than Π.

If E (resp. F) is normable, Γ_E (resp. Γ_F) can be replaced by a singleton, and it follows in both cases that (2) implies (1), hence $\Pi = \Delta$.

0.6.3. Remark. In the general case where neither E nor F is normable, Π may be strictly coarser than Δ, even if both spaces are metrizable. For example on $\mathcal{L}(\mathbb{R}^{\mathbb{N}},\mathbb{R}^{\mathbb{N}})$ one has $\Pi \neq \Delta$.

However $\mathcal{L}^n_\Pi(E,F)$ and $\mathcal{L}^n_\Delta(E,F)$ have always the same convergent sequences.

0.6.4. Corollary. The evaluation mapping

$$ev : \mathcal{L}^n_\Delta(E,F)\times E^n \to F$$

is continuous.

0.6.5. Corollary. If E is normable, then $\Delta = \Pi = \Lambda_{qb} = \Lambda_b$ is an l.c. topology on $\mathcal{L}^n(E,F)$.

If E is not normable, $F \neq \{0\}$ and $n \neq 0$, then Δ is no topology and there is no compatible topology finer than Δ on $\mathcal{L}^n(E,F)$.

0.6.6. Theorem. Let E and F be two l.c.s. and $n \in \mathbb{N}$. Assume that $H = H_1\times\ldots\times H_m$ is a finite product of topological v.s. For an m-linear mapping $w : H \to \mathcal{L}^n(E,F)$ the following statements are equivalent:

(i) $w : H \to \mathcal{L}^n_\Delta(E,F)$ is continuous.

(ii) $w : H \to \mathcal{L}^n_\Pi(E,F)$ is continuous.

(iii) $w : H \to \mathcal{L}^n_{qb}(E,F)$ is continuous.

(iv) $w : H \to \mathcal{L}^n_c(E,F)$ is continuous.

Proof. We only have to show that (iv) implies (i). Assume that w satisfies (iv). By 0.3.7. (2') there is a $\phi \in \Phi$ and for every $\beta \in \Gamma_F$ there are 0-neighbourhoods Z_i' in H_i , $1 \leq i \leq m$, and $\mu > 0$ such that

$$\sup\{|wz_1\ldots z_m|_{\beta,\phi(\beta)}\,|\,z_i \in Z_i', 1\leq i\leq m\} \leq \mu \ .$$

Given $\varepsilon > 0$, if we choose $Z_1 := \varepsilon\cdot\mu^{-1}\cdot Z_1'$ and $Z_j := Z_j'$

for $2 \leq j \leq m$, we have

$$|wz_1 \ldots z_m|_{\beta,\phi(\beta)} \leq \varepsilon$$

for all $z_i \in Z_i$, $1 \leq i \leq m$. As $Z_1 \times \ldots \times Z_m$ is absorbing in H we have $w(H) \subset \mathscr{L}^n_\phi(E,F)$ and w maps H continuously into $\mathscr{L}^n_\phi(E,F)$, hence also continuously into $\mathscr{L}^n_\Delta(E,F)$.

0.6.7. <u>Corollary</u>. Let E and F be two l.c.s. and $n \in \mathbb{N}$. Let Λ denote any convergence structure, compatible or not, on $\mathscr{L}^n(E,F)$ which is finer than Λ_c and coarser than Δ. Then for every $m \in \mathbb{N}$ the canonical linear isomorphism

$$\Theta^{m,n} : L^m(E,L^n(E,F)) \to L^{m+n}(E,F)$$

induces an isomorphism

$$\Theta^{m,n} : \mathscr{L}^m(E,\mathscr{L}^n_\Lambda(E,F)) \to \mathscr{L}^{m+n}(E,F)$$

of v.s.

Proof. According to 0.3.4. $\Theta^{m,n}$ maps the v.s. $\mathscr{L}^m(E,\mathscr{L}^n_c(E,F))$ isomorphically onto the v.s. $\mathscr{L}^{m+n}(E,F)$. As a consequence of 0.6.6. we have

$$\mathscr{L}^m(E,\mathscr{L}^n_\Lambda(E,F)) = \mathscr{L}^m(E,\mathscr{L}^n_c(E,F)) .$$

0.6.8. <u>Remarks</u>. (1) The statement of 0.6.7. is especially important if we apply it to $\Lambda = \Delta$, Π, Λ_{qb}, Λ_c respectively. In these cases we shall, in the forthcoming chapters, always, if not otherwise stated, identify $\mathscr{L}^m(E,\mathscr{L}^n_\Lambda(E,F))$ and $\mathscr{L}^{m+n}(E,F)$

qua v.s. by means of $\Theta^{m,n}$.

(2) If $\Lambda = \Lambda_c$ we need only require that E and F are c.v.s. Then by 0.3.4. $\Theta^{m,n}$ is even an isomorphism of c.v.s. from $\mathcal{L}_c^m(E,\mathcal{L}_c^n(E,F))$ onto $\mathcal{L}_c^{m+n}(E,F)$.

(3) In case $\Lambda = \Lambda_{qb}$ an analogous statement holds if E is an l.c.s. and F any c.v.s. (cf. 0.4.11.)

(4) If $\Lambda = \Pi$ or $\Lambda = \Delta$ the spaces E and F both have to be l.c.s. and $\mathcal{L}_\Lambda^m(E,\mathcal{L}_\Lambda^n(E,F))$ has no a priori meaning, unless E is normable or $F = \{0\}$ or $n = 0$. However, we may define Λ on $\mathcal{L}^m(E,\mathcal{L}_\Lambda^n(E,F))$ by the requirement that $\Theta^{m,n}$ becomes an isomorphism of c.v.s. onto $\mathcal{L}_\Lambda^{m+n}(E,F)$.

0.6.9. <u>Theorem</u>. Let E , F and G be three l.c.s. Assume that $n \in \mathbb{N}$ and that $m_1,\dots,m_n$ are elements of $\mathbb{N}$ with sum m . Then the canonical (n+1)-linear mapping

$$\pi : \mathcal{L}_\Delta^n(F,G) \times \prod_{1 \leq i \leq n} \mathcal{L}_\Delta^{m_i}(E,F) \to \mathcal{L}_\Delta^m(E,G)$$

defined by

$$\pi(v,(u_1,\dots,u_n)) := v \circ (u_1 \times \dots \times u_n) ,$$

is continuous.

Proof. Let us denote, as usual, by Γ_E , Γ_F , Γ_G the directed sets of all continuous semi-norms in E , F , G respectively. Now for every choice of $\phi \in \Phi := (\Gamma_E)^{\Gamma_F}$ and $\psi \in (\Gamma_F)^{\Gamma_G}$, thus $\phi \circ \psi \in (\Gamma_E)^{\Gamma_G}$, the l.c.s.

$$\mathcal{L}^{n}_{\psi}(F,G)\times\prod_{1\le i\le n}\mathcal{L}^{m_i}_{\phi}(E,F)$$

is mapped continuously into the l.c.s. $\mathcal{L}^{m}_{\phi\circ\psi}(E,G)$ and hence into $\mathcal{L}^{m}_{\Delta}(E,G)$ by π . But as the domain of π is the inductive limit (in the category of convergence spaces) of these products of l.c.s., this means that π is continuous. The main argument of this proof is based on the fact that

$$|v\circ(u_1\times\ldots\times u_n)|_{\gamma,\alpha}\le|v|_{\gamma,\beta}\cdot|u_1|_{\beta,\alpha}\cdot\ldots\cdot|u_n|_{\beta,\alpha}$$

for all $u_i\in\mathcal{L}^{m_i}(E,F)$, $1\le i\le n$, $v\in\mathcal{L}^{n}(F,G)$ and all $\alpha\in\Gamma_E$, $\beta\in\Gamma_F$, $\gamma\in\Gamma_G$.

0.7. The convergence structure Θ on $\mathcal{L}^n(E,F)$

The convergence structure Θ to be introduced on $\mathcal{L}^n(E,F)$ in this section will be always compatible with addition; scalar multiplication however will be continuous, simultaneously in both variables, only in special cases. A certain notion of differentiability will be based on Θ in the forthcoming chapters.

0.7.1. Proposition. Let E and F be two l.c.s. For every $n\in\mathbb{N}$ there exists a unique separated convergence structure Θ on the v.s. $\mathcal{L}^n(E,F)$ which is invariant with respect to translations and such that for every filter $\mathcal{F}$ on $\mathcal{L}^n(E,F)$ we have

$$\mathcal{F}\in\Theta(0)\quad\text{iff}\quad(\exists\alpha\in\Gamma_E)(\forall\beta\in\Gamma_F)(\forall\varepsilon>0)(\exists Q\in\mathcal{F})\ \sup_{u\in Q}|u|_{\beta,\alpha}\le\varepsilon$$

The convergence space

$$\mathcal{L}^n_\Theta(E,F) := (\mathcal{L}^n(E,F),\Theta)$$

has the following properties:

I. Addition $(u,v) \longmapsto u+v$ is a continuous (linear) mapping from $\mathcal{L}^n_\Theta(E,F)\times\mathcal{L}^n_\Theta(E,F)$ onto $\mathcal{L}^n_\Theta(E,F)$.

II. For every $\lambda \in \mathbb{R}$, multiplication by λ is a continuous (linear) mapping $u \longmapsto \lambda\cdot u$ from $\mathcal{L}^n_\Theta(E,F)$ into $\mathcal{L}^n_\Theta(E,F)$.

III. Scalar multiplication $(\lambda,u) \longmapsto \lambda\cdot u$, as a (bilinear) mapping from $\mathbb{R}\times\mathcal{L}^n_\Theta(E,F)$ onto $\mathcal{L}^n_\Theta(E,F)$ is continuous at $(0,0)$.

Proof. It is easy to verify that the set $\Theta(0)$ of filters on $\mathcal{L}^n(E,F)$, as defined in 0.7.1., has the following properties: (a) $\mathcal{F}_1 \in \Theta(0)$ and $\mathcal{F}_2 \in \Theta(0)$ imply $\mathcal{F}_1 \wedge \mathcal{F}_2 \in \Theta(0)$; (b) $\mathcal{F} \in \Theta(0)$ and $\mathcal{G}$ a filter on $\mathcal{L}^n(E,F)$, finer than $\mathcal{F}$, imply $\mathcal{G} \in \Theta(0)$; (c) $[0] \in \Theta(u)$; (d) $\mathcal{F}_1 \in \Theta(0)$ and $\mathcal{F}_2 \in \Theta(0)$ imply $\mathcal{F}_1+\mathcal{F}_2 \in \Theta(0)$; (e) $\lambda \in \mathbb{R}$ and $\mathcal{F} \in \Theta(0)$ imply $\lambda\cdot\mathcal{F} \in \Theta(0)$; (f) $\mathcal{F} \in \Theta(0)$ implies $\mathbb{V}\cdot\mathcal{F} \in \Theta(0)$; (g) $u \in \mathcal{L}^n(E,F)$, $[u] \in \Theta(0)$ imply $u = 0$.

If, for every $u \in \mathcal{L}^n(E,F)$ and every filter $\mathcal{F}$ on $\mathcal{L}^n(E,F)$ we define $\mathcal{F} \in \Theta(u)$ by $\mathcal{F}-u\in\Theta(0)$, then, as a consequence of (a), (b), (c) the assignment $u \longmapsto \Theta(u)$ is a convergence structure on $\mathcal{L}^n(E,F)$, invariant with respect to translations. The properties (d), (e), (f) correspond to the statements (I), (II), (III) respectively. Finally, taking (I) into account, (g) expresses separatedness of Θ .

0.7.2. Remark. We do not claim that $\mathbb{V}\cdot u \in \Theta(0)$ for every $u \in \mathcal{L}^n(E,F)$ and consequently neither that scalar multiplication be continuous everywhere; therefore $\mathcal{L}^n_\Theta(E,F)$ will be an abelian convergence group (with respect to addition), but not a c.v.s., except in special cases, e.g. if E or F is normable.

As a matter of fact, $\mathbb{V}\cdot u \in \Theta(0)$ iff $u \in \mathcal{L}^n(E,F)$ has the property that there exists an $\alpha \in \Gamma_E$ such that $|u|_{\beta,\alpha} < \infty$ for every $\beta \in \Gamma_F$. As a consequence the restriction of Θ to the linear subspace

$$\tilde{\mathcal{L}}^n(E,F) := \{u \in \mathcal{L}^n(E,F) \,|\, (\exists \alpha \in \Gamma_E)(\forall \beta \in \Gamma_F)\, |u|_{\beta,\alpha} < \infty\}$$

of $\mathcal{L}^n(E,F)$ is compatible with scalar multiplication, thus yielding a c.v.s. $\tilde{\mathcal{L}}^n_\Theta(E,F)$.

0.7.3. Proposition. Let E and F be l.c.s. For each $n \in \mathbb{N}$ the convergence structure Θ on $\mathcal{L}^n(E,F)$ is finer than Π. If E or F is normable, then $\Theta = \Pi = \Delta$.

Proof. From the definitions in 0.7.1. and 0.5.1. it is obvious that $\Theta(0) \subset \Pi(0)$, hence Θ is finer than Π.

If E or F is normable, Γ_E respectively Γ_F can be reduced to a singleton. In each of these cases the conditions for a filter on $\mathcal{L}^n(E,F)$ to be an element of $\Theta(0)$ or of $\Pi(0)$ coincide; hence $\Theta = \Pi$, and according to 0.6.2., $\Pi = \Delta$.

0.7.4. Corollary. (1) If E is normable, then

$$\mathcal{L}^n_\Theta(E,F) = \tilde{\mathcal{L}}^n_\Theta(E,F) = \mathcal{L}^n_\Delta(E,F) = \mathcal{L}^n_\Pi(E,F) = \mathcal{L}^n_{qb}(E,F) = \mathcal{L}^n_b(E,F)$$

is an l.c.s.

(2) If F is normable, then

$$\mathcal{L}^n_\Theta(E,F) = \tilde{\mathcal{L}}^n_\Theta(E,F) = \mathcal{L}^n_\Delta(E,F) = \mathcal{L}^n_\Pi(E,F)$$

is a c.v.s.

Proof. One only has to observe that in each of these cases $\mathcal{L}^n(E,F) = \tilde{\mathcal{L}}^n(E,F)$.

0.7.5. Corollary. Let E and F be two l.c.s. and $n \in \mathbb{N}$. The evaluation mapping

$$ev : \mathcal{L}^n_\Theta(E,F) \times E^n \to F$$

is continuous.

0.7.6. Corollary. Let E and F be l.c.s. For every $(m,n) \in \mathbb{N}\times\mathbb{N}$ the canonical linear isomorphism

$$\Theta^{m,n} : L^m(E,L^n(E,F)) \to L^{m+n}(E,F)$$

induces a linear isomorphism of the v.s. $\mathcal{L}^m(E,\mathcal{L}^n_\Theta(E,F))$ onto a linear subspace of the v.s. $\mathcal{L}^{m+n}(E,F)$.

0.7.7. Theorem. Let E , F and G be three l.c.s. Assume that $n \in \mathbb{N}$ and that $m_1,\ldots,m_n$ are elements of $\mathbb{N}$ with sum m . Then the canonical $(n+1)$-linear mapping

$$\pi : \mathscr{L}^n_\Theta(F,G)\times\prod_{1\leq i\leq n}\mathscr{L}^{m_i}_\Theta(E,F) \to \mathscr{L}^m_\Theta(E,G) \ ,$$

defined by $\pi(v,(u_1,\ldots,u_n)) := v\circ(u_1\times\ldots\times u_n)$, is continuous.

Proof. For the sake of simplicity we carry out the proof of 0.7.7. in the special case $n = 1$ and $m = m_1 = 1$. The general case can be dealt with using essentially the same arguments.

We thus have to prove that composition $(v,u)\longmapsto v\circ u$, as a bilinear map

$$\pi : \mathscr{L}_\Theta(F,G)\times\mathscr{L}_\Theta(E,F) \to \mathscr{L}_\Theta(E,G) \ ,$$

is continuous at an arbitrary point (v_o,u_o) of its domain. To this purpose we assume that $\mathscr{F}$ and $\mathscr{G}$ are filters convergent to 0 in $\mathscr{L}_\Theta(E,F)$ respectively $\mathscr{L}_\Theta(F,G)$, and we have to show that the filter $(v_o+\mathscr{G})\circ(u_o+\mathscr{F}) - v_o\circ u_o$ which is finer than $v_o\circ\mathscr{F} + \mathscr{G}\circ u_o + \mathscr{G}\circ\mathscr{F}$ converges to 0 in $\mathscr{L}_\Theta(E,G)$. Taking into account 0.7.1.(I), it is sufficient to prove that each of the filters $v_o\circ\mathscr{F}$, $\mathscr{G}\circ u_o$, $\mathscr{G}\circ\mathscr{F}$ on $\mathscr{L}_\Theta(E,G)$ converges to 0 .

According to the definition of Θ we may fix $\alpha_o\in\Gamma_E$ and $\beta_o\in\Gamma_F$ such that

$$(\forall\beta\in\Gamma_F)(\forall\varepsilon>0)(\exists Q\in\mathscr{F}) \qquad \sup_{u\in Q}|u|_{\beta,\alpha_o}\leq\varepsilon \ ;$$

$$(\forall\gamma\in\Gamma_G)(\forall\varepsilon>0)(\exists R\in\mathscr{G}) \qquad \sup_{v\in R}|v|_{\gamma,\beta_o}\leq\varepsilon \ .$$

Let $\gamma \in \Gamma_G$ and $\varepsilon > 0$ be arbitrary.

a) We first choose $\beta \in \Gamma_F$ such that $\nu := |v_0|_{\gamma,\beta} < \infty$, and then we choose $Q \in \mathcal{F}$ such that $|u|_{\beta,\alpha_0} \leq \varepsilon\cdot\nu^{-1}$ for all $u \in Q$. It follows

$$\sup_{u\in Q} |v_0 \circ u|_{\gamma,\alpha_0} \leq \nu\cdot\sup_{u\in Q} |u|_{\beta,\alpha_0} \leq \varepsilon .$$

This proves that $v_0 \circ \mathcal{F}$ converges to 0 in $\mathcal{L}_\Theta(E,G)$.

b) We fix $\alpha_1 \in \Gamma_E$ such that $\mu := |u_0|_{\beta_0,\alpha_1} < \infty$. We now can choose $R \in \mathcal{G}$ such that $|v|_{\gamma,\beta_0} \leq \varepsilon\cdot\mu^{-1}$ for all $v \in R$. We get

$$\sup_{v\in R} |v \circ u_0|_{\gamma,\alpha_1} \leq \sup_{v\in R} |v|_{\gamma,\beta_0}\cdot\mu \leq \varepsilon ,$$

which shows that $\mathcal{G} \circ u_0$ converges to 0 in $\mathcal{L}_\Theta(E,G)$.

c) If we choose $Q \in \mathcal{F}$ and $R \in \mathcal{G}$ such that

$$\sup_{u\in Q} |u|_{\beta_0,\alpha_0} \leq \varepsilon \quad\text{and}\quad \sup_{v\in R} |v|_{\gamma,\beta_0} \leq 1 ,$$

we have

$$\sup_{u\in Q}\sup_{v\in R} |v\circ u|_{\gamma,\alpha_0} \leq \sup_{v\in R} |v|_{\gamma,\beta_0}\cdot\sup_{u\in Q} |u|_{\beta_0,\alpha_0} \leq \varepsilon .$$

This shows that $\mathcal{G}\circ\mathcal{F}$ converges to 0 in $\mathcal{L}_\Theta(E,G)$ and thereby completes the proof of 0.7.7.

1. CONTINUOUSLY DIFFERENTIABLE FUNCTIONS

1.0. General concept: Functions of class C^1_Λ

Unless otherwise stated, in this chapter E and F will always denote l.c.s. with Γ_E, Γ_F the directed sets of all continuous semi-norms in E, F respectively; and X will denote an open subset of E. We are going to introduce a general concept of continuous differentiability for continuous functions $f : X \to F$ by assuming that the directional derivative $Df(x)h$ of f exists at each point $x \in X$ in each direction $h \in E$, that $h \longmapsto Df(x)h$ is a continuous linear mapping from E into F (i.e. f is differentiable in the sense of Gâteaux-Lévy) and finally that the assignment $x \longmapsto Df(x)$, as a function from X into $\mathcal{L}(E,F)$, is continuous with respect to some given convergence structure (or topology) Λ on $\mathcal{L}(E,F)$. Our main attention will be paid to the cases where Λ coincides with one of the structures introduced in Chapter 0. We thus start with the following formal definition:

1.0.0. <u>Definition</u>. Let Λ be a convergence structure on $\mathcal{L}(E,F)$. A continuous function $f : X \to F$ is said to be <u>differentiable of class</u> C^1_Λ if there exists a continuous mapping

$$Df : X \to \mathcal{L}_\Lambda(E,F) ,$$

called the <u>derivative of</u> f, such that for every $(x,h) \in X \times E$

$$\lim_{t \to 0 \in \mathbb{R}} \frac{1}{t}(f(x+th)-f(x)) = Df(x)h .$$

Let $C^0(X,F)$, $C^1_\Lambda(X,F)$ denote the sets of functions from X to F which are continuous, respectively differentiable of class C^1_Λ . Likewise $C^0(X,\mathcal{L}_\Lambda(E,F))$ will denote the set of all continuous mappings from X into the convergence space $\mathcal{L}_\Lambda(E,F)$. Clearly $C^0(X,F)$ is a linear subspace of the v.s. F^X , and $C^0(X,\mathcal{L}_\Lambda(E,F))$ is a linear subspace of the v.s. $(\mathcal{L}(E,F))^X$, provided addition and multiplication by each $\lambda \in \mathbb{R}$ are continuous operations in $\mathcal{L}_\Lambda(E,F)$. The verification of the following statement is straightforward.

1.0.1. <u>Proposition</u>. Assume that Λ is a convergence structure on $\mathcal{L}(E,F)$ such that addition and multiplication by each $\lambda \in \mathbb{R}$ are continuous operations in $\mathcal{L}_\Lambda(E,F)$. Then $C^1_\Lambda(X,F)$ is a linear subspace of the v.s. F^X and differentiation $D : C^1_\Lambda(X,F) \to C^0(X,\mathcal{L}_\Lambda(E,F))$ is a linear map.

In the case where Λ is equal to one of the convergence structures Θ , Δ , Π , Λ_{qb} , Λ_c or one of the l.c. topologies $\Lambda_{\mathfrak{S}}$ ($\mathfrak{S}$ a cover of E by bounded sets) we write C^1_Θ , C^1_Δ , C^1_Π , C^1_{qb} , C^1_c , $C^1_{\mathfrak{S}}$ respectively for C^1_Λ . We also use C^1_b , C^1_{pk} , C^1_k , C^1_s for $C^1_{\mathfrak{S}}$ if $\Lambda_{\mathfrak{S}}$ is equal to Λ_b , Λ_{pk} , Λ_k , Λ_s respectively.

We now take into account the relations which we have established in Chapter 0 between the various convergence structures on $\mathcal{L}(E,F)$. From each of those results we can directly deduce a corresponding assertion on the notions of continuous differentiability based on these structures. In order to summarize these assertions in a clear and concise form we shall use the following notations and symbols. Given a continuous function $f : X \to F$ let "C^1_Λ" mean "$f \in C^1_\Lambda(X,F)$" where Λ is a convergence structure on $\mathcal{L}(E,F)$. A normal arrow " $\to$ " will denote unconditional implication, i.e. holding for any choice of E , F and X , whereas a dashed arrow " $\dashrightarrow$ ", together with a condition on E or F , denotes an implication which holds whenever the indicated condition is fulfilled. The abbreviations "metr." , "norm." , "compl." , "barr." , "Sch." mean respectively "metrizable", "normable", "complete", "barrelled", "Schwartz".

1.0.2. Theorem. Let E and F be two l.c.s. and let X be an open set in E . For continuous functions $f : X \to F$ the implications indicated in Table 1 hold.

Proof. Let Λ , Λ' be any convergence structures on $\mathcal{L}(E,F)$. If we know that

$$C^0(X,\mathcal{L}_\Lambda(E,F)) \subset C^0(X,\mathcal{L}_{\Lambda'}(E,F)) ,$$

then from definition 1.0.0. we conclude that $C^1_\Lambda(X,F) \subset C^1_{\Lambda'}(X,F)$. Let us apply this simple general remark to those assertions in chapter 0 which concern the relationship between the various structures on $\mathcal{L}(E,F)$. The result will be the set of implications between the corresponding notion of continuous differentiability indicated in Table 1.

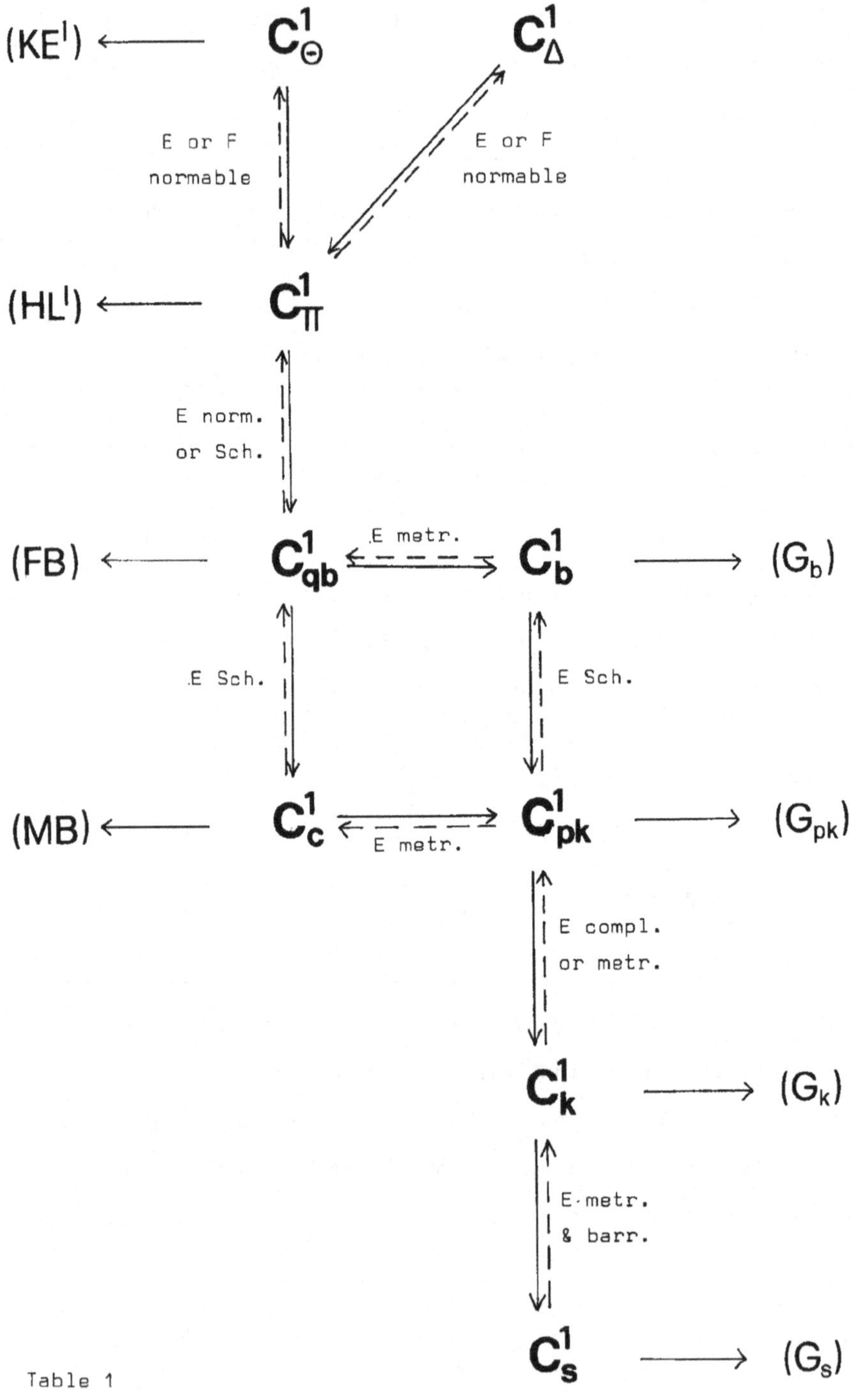
(KE^I)
C¹_Θ
C¹_Δ
E or F normable
E or F normable
(HL^I)
C¹_π
E norm. or Sch.
(FB)
C¹_qb
E metr.
C¹_b
(G_b)
E Sch.
E Sch.
(MB)
C¹_c
E metr.
C¹_pk
(G_pk)
E compl. or metr.
C¹_k
(G_k)
E metr. & barr.
C¹_s
(G_s)

Table 1

We give a few examples. 1) As by 0.4.4. and 0.4.8. Λ_{qb} is always finer than Λ_b and Λ_c , we have the unconditional implications $C^1_{qb} \to C^1_b$, and $C^1_{qb} \to C^1_c$. 2) If E , and hence X , is metrizable, by 0.3.9. resp. 0.4.7. we have $C^0(X,\mathscr{L}_k(E,F)) = C^0(X,\mathscr{L}_c(E,F))$ and $C^0(X,\mathscr{L}_b(E,F)) = C^0(X,\mathscr{L}_{qb}(E,F))$ therefore $C^1_k(X,F) = C^1_c(X,F)$ and $C^1_b(X,F) = C^1_{qb}(X,F)$.
3) If E is a Schwartz space, by 0.5.7., $\mathscr{L}_c(E,F) = \mathscr{L}_\Pi(E,F)$ and therefore $C^1_c(X,F) = C^1_\Pi(X,F)$. We leave the remaining cases to the reader.

1.0.3. <u>Corollary</u>. Let E be a normable, F an arbitrary l.c.s. and X an open set in E . Then

(1) C^1_Θ , C^1_Δ , C^1_Π , C^1_{qb} , C^1_b are equivalent.

(2) C^1_c , C^1_{pk} , C^1_k are equivalent.

If moreover E is complete, i.e. a Banach space, then all of the statements C^1_c , C^1_{pk} , C^1_k , C^1_s are equivalent.

Thus, for functions from an open subset X of a Banach space E to any l.c.s. F , in our setting, we only have two notions of continuous differentiability which we could call continuous Fréchet differentiability and continuous Gâteaux-Lévy differentiability respectively. As a consequence of 0.5.5. and 1.2.8., if also F is normable, C^1_b turns out to mean continuous Fréchet differentiability in the classical sense of [20]. It can be shown that there exist Banach spaces E , F and functions $f:E \to F$ which are of class C^1_s but not of class C^1_b (cf. [67], § 1.2).

1.0.4. Corollary. Let E be a Fréchet space, F an arbitrary l.c.s. and X an open set in E. Then

(1) C^1_{qb} and C^1_b are equivalent.

(2) C^1_c, C^1_{pk}, C^1_k, C^1_s are equivalent.

1.0.5. Corollary. Let E be a Fréchet-Schwartz space, F an arbitrary l.c.s. and X an open set in E. Then all the statements C^1_Π, C^1_{qb}, C^1_c, C^1_b, C^1_{pk}, C^1_k, C^1_s are equivalent.

1.0.6. Corollary. Let E be a finite-dimensional and F an arbitrary l.c.s., or let E be a Fréchet-Schwartz space and F a normable l.c.s. Let X be an open subset of E. Then for every function $f : X \to F$ all the statements of the form C^1_Λ which occur in Table 1 are equivalent.

Thus, if E has finite dimension and F is an arbitrary l.c.s. or if E is a Fréchet-Schwartz space and F a normable l.c.s., within the frame of our theory, there is a unique notion of continuous differentiability for functions from an open set X in E to F, which we shall call natural, as it does not depend on any convergence structure or topology on $\mathcal{L}(E,F)$, at least as far as our view is restricted to the structures introduced in Chapter 0. Functions $f : X \to F$ having this property will consequently simply be referred to as continuously differentiable or differentiable of class C^1, and the v.s. of all such functions will simply be denoted by $C^1(X,F)$.

If we disregard the convergence structures Θ and Δ, i.e. if we are only concerned with structures on $\mathcal{L}(E,F)$ which are coarser than Π and finer than Λ_s, we may, on the ground of 1.0.5., apply the terminology and notation just introduced even to the more general case that E is a Fréchet-Schwartz space while F remains arbitrary.

1.0.7. Remark. Let J be an open set in $\mathbb{R}$ and F an l.c.s. It is immediate that a function $f : J \to F$ is differentiable of class C^1 in the sense above iff it is continuously differentiable in the usual sense, i.e. of class C^1 according to definition A.1.1. Assume that $f \in C^1(J,F)$. Then $f'(t) = Df(t)(1)$, i.e. $Df = f'$ if $\mathcal{L}(\mathbb{R},F)$ is identified to F by means of the canonical isomorphism $u \longmapsto u(1)$.

1.1. Auxiliary formulae

The following notation will be useful in the sequel:

If a and b are points in a v.s. E, we denote by

$$[a,b] := \{a+t(b-a) \mid t \in [0,1] \subset \mathbb{R}\}$$

the closed segment in E joining a to b.

In this whole section E and F will be two l.c.s. and X an open set in E.

1.1.0. Definitions. Let $f : X \to F$ be of class C^1_Λ with respect to some convergence structure Λ on $\mathcal{L}(E,F)$. For each $x \in X$ and each $h \in E$ such that $x+h \in X$ the <u>remainder of</u> f <u>at</u> x <u>in the direction</u> h is defined by

$$Rf_x(h) = Rf(x,h) := f(x+h)-f(x)-Df(x)h \ .$$

The <u>remainder</u> Rf <u>of</u> f may be viewed as a mapping

$$Rf : \Omega_X := \{(x,h) \in X \times E \mid x+h \in X\} \to F \ .$$

We shall especially be interested in the behaviour of Rf as a function of its second argument, i.e., for any fixed $x \in X$, the mapping

$$Rf_x : X-x \to F \ ,$$

called the <u>remainder of</u> f <u>at</u> x.

In some investigations on remainders the following expression, introduced by A. Bastiani (cf. [11]) and also by Frölicher and Bucher (cf. [31]) will play a major rôle: For $x \in X$ fixed define

$$\Theta Rf_x : \Omega_x := \{(t,h) \in \mathbb{R} \times E \mid x+th \in X\} \to F \ ,$$

by

$$\Theta Rf_x(t,h) := t^{-1} Rf_x(th) \ , \text{ if } \ t \neq 0 \ ;$$
$$\Theta Rf_x(0,h) := 0 \ .$$

As an immediate consequence of definition 1.0.0. we get

1.1.1. Lemma. If $f : X \to F$ is of class C^1_s , then for every $(x,h) \in E\times E$ the function of a real variable

$$\phi : J_{x,h} := \{t \in \mathbb{R} \mid x+th \in X\} \to F ,$$

defined by

$$\phi(t) := f(x+th) ,$$

is of class C^1 in the usual sense (cf. A.1.1.) and its derivative $\phi' : J_{x,h} \to F$ is given by

$$\phi'(t) = Df(x+th)h .$$

From 1.1.1., together with 1.1.0., we get

1.1.2. Lemma. Under the same assumptions as in 1.1.1. the function $\psi : J_{x,h} \to F$, defined by

$$\psi(t) := Rf_x(th) ,$$

is of class C^1 in the usual sense (cf. A.1.1.) and we have

$$\psi'(t) = (Df(x+th)-Df(x))h$$

for every $t \in J_{x,h}$.

1.1.3. Proposition. Let $f : X \to F$ be of class C^1_s . For every $(x,h) \in X\times E$ such that $[x,x+h] \subset X$ we have the following representations by Riemann integrals:

(1) $f(x+h)-f(x) = \int_0^1 Df(x+th)h \, dt$;

(2) $Rf_x(h) = \int_0^1 (Df(x+th)-Df(x))h \, dt$.

For every $(x,h) \in X \times E$ and every $t \in \mathbb{R}$, $t \neq 0$, such that $[x,x+th] \subset X$, we have

(3) $\Theta Rf_x(t,h) = \frac{1}{t} \int_0^t (Df(x+\tau h)-Df(x))h \, d\tau$.

Proof. (1) and (2) are obtained when we apply A.3.3. to 1.1.1. and 1.1.2. respectively; (3) follows from (2) by a trivial computation.

1.1.4. Corollary. Under the same assumptions as in 1.1.3., for every $\beta \in \Gamma_F$, we have the inequalities:

(1) $|f(x+h)-f(x)|_\beta \leq \sup_{0 \leq t \leq 1} |Df(x+th)h|_\beta$;

(2) $|Rf_x(h)|_\beta \leq \sup_{0 \leq t \leq 1} |(Df(x+th)-Df(x))h|_\beta$;

(3) $|\Theta Rf_x(t,h)|_\beta \leq \sup_{\tau \in [0,t]} |(Df(x+\tau h)-Df(x))h|_\beta$.

Proof. Application of A.2.7. to 1.1.3.

1.1.5. Remark. The assertions in 1.1.1., 1.1.2., 1.1.3. and 1.1.4. are still true if we only assume $f : X \to F$ to have the following properties:

There exists a continuous mapping $Df : X \to F^E$ (with respect to Λ_s in F^E), such that the limit

$$\lim_{t \to 0} \frac{1}{t}(f(x+th)-f(x)) = Df(x)(h)$$

exists for each $(x,h) \in X \times E$; neither linearity nor continuity of $Df(x) : E \to F$ is necessary.

As a matter of fact, by a general mean value theorem of differential calculus ("théorème des accroissements finis", cf. [15]), for the validity of the inequalities (1), (2), (3) in 1.1.4. even the continuity of $Df : X \to F^E$ is redundant.

1.2. Properties of remainders

1.2.0. General remarks. In this section, as in the whole chapter, E and F denote two l.c.s. and X an open set in E. Given a function $f : X \to F$ which is differentiable of class C^1_Λ for some topology or convergence structure Λ on $\mathcal{L}(E,F)$, we shall investigate the behaviour of the remainder $Rf_x : X-x \to F$ of f at a point $x \in X$; especially, of course, we shall be interested in the value of $Rf_x(h)$ for "small" h. As a result we shall receive, depending on the choice of Λ, some of the most important conditions which, during the last few decades, have been imposed on the remainders in the various attempts to define differentiability of f at a fixed point x. We refer here to the excellent survey article by Averbukh and Smolyanov [6] and to their appendix to the Russian translation of Frölicher and Bucher's book [31] (cf. [8], Russian).

The choice of adequate symbols to denote the various "remainder conditions" causes some difficulties because they have frequently been changed in the short history of differential calculus in non-normable spaces and notation varies from author to author; no universally accepted convention exists so far. In the case of remainders of "Gâteaux type", corresponding to $\mathfrak{S}$-topologies on $\mathcal{L}(E,F)$, we shall use, according to our 1964 article [36], the symbol $(G_{\mathfrak{S}})$. For the remainders which result when $\mathcal{L}(E,F)$ is endowed with Λ_c , Λ_{qb} , Π , Θ resp., we have adopted here the notations used in [8], namely (MB) , (FB) , (HL') , (KE') resp.; the last two result from applying essentially the same modification to two stronger remainder conditions (HL) , (KE) respectively.

In Table 1 we are indicating the remainder properties which correspond to the various kinds of differentiability of class C^1_Λ .

1.2.1. <u>Theorem</u>. Let $\mathfrak{S}$ be a cover of E by bounded sets. Assume that $f : X \to F$ is of class $C^1_{\mathfrak{S}}$. Then for each $x \in X$ the remainder $Rf_x : X-x \to F$ satisfies the condition

$$(G_{\mathfrak{S}}) \qquad \lim_{t\to 0} \frac{1}{t} Rf_x(th) = \lim_{t\to 0} \Theta Rf_x(t,h) = 0 ,$$

uniformly with respect to h in each set $S \in \mathfrak{S}$.

Proof. Given $\beta \in \Gamma_F$, $\varepsilon > 0$ and $S \in \mathfrak{S}$, according to the hypothesis on Df , there exists a circled neighbourhood

U of 0 in E, such that $h \in S$, $h' \in U \cap (X-x)$ imply

$$|(Df(x+h') - Df(x))h|_\beta \leq \varepsilon .$$

We choose $\delta > 0$ such that $tS \subset U$ if $|t| \leq \delta$. From 1.1.4.(3) we get

$$|\Theta Rf_x(t,h)|_\beta \leq \varepsilon \quad \text{whenever} \quad h \in S , \ |t| \leq \delta .$$

1.2.2. Remarks. (1) The idea of defining a function $f : X \to F$ to be "differentiable with respect to a system $\mathfrak{S}$ of sets" at a point $x \in X$ by imposing $(G_{\mathfrak{S}})$ on the remainder of f at x, appears already, though in a wider setting, in J. Gil de Lamadrid's thesis 1955 (cf. [33] and [34]) as well as in the work of J. Sebastião e Silva 1956 (cf. [52]). The same concept is used in [36].

(2) If for $\mathfrak{S}$ we choose one of the special collections $\mathfrak{S}_s$, $\mathfrak{S}_k$, $\mathfrak{S}_{pk}$, $\mathfrak{S}_b$ of sets (cf. Sect. 0.1.), we shall denote the corresponding remainder property $(G_{\mathfrak{S}})$ by $(G_s),\ldots,(G_b)$ respectively.

(3) The property (G_s) just means that f is differentiable at x in the classical sense of Gâteaux-Lévy. But this is what, by definition 1.0.0., has been required a priori! Thus, for C^1_s the statement of 1.2.1. is vacuous.

(4) The other extreme, the condition (G_b) of "bounded differentiability" of f at x, is, applied to normed (or normable) spaces E, F, the statement that f is differentiable at x in the classical sense of Fréchet.

1.2.3. <u>Theorem</u>. Assume that $f : X \to F$ is of class C^1_c . Then for every $x \in X$ the remainder $Rf_x : X-x \to F$ satisfies the condition

$$\text{(MB)} \qquad (\forall h_o \in E) \lim_{(t,h)\to(0,h_o)} \Theta Rf_x(t,h) = 0 .$$

Proof. By the hypothesis on Df , for every $h_o \in E$, every $\beta \in \Gamma_F$ and every $\varepsilon > 0$ there exists a circled neighbourhood U of 0 in E such that

$$|(Df(x+h')-Df(x))h|_\beta \leq \varepsilon \quad \text{if} \quad h-h_o \in U \quad \text{and} \quad h' \in U .$$

We can choose $\delta > 0$ such that $\delta\cdot(h_o+U) \subset U$. We then have

$$|(Df(x+th)-Df(x))h|_\beta \leq \varepsilon \quad \text{whenever} \quad |t|\leq\delta \quad \text{and} \quad h-h_o \in U .$$

Now 1.1.4. (3) yields the desired result.

1.2.4. <u>Remark</u>. The condition (MB) , also denoted by this symbol in Averbukh and Smolyanov's survey article [6] and in [8], is the condition for differentiability of f at x in Andrée Bastiani's theory of differentiation (cf. [10] and [11]) in arbitrary c.v.s. It is equivalent to an earlier definition of remainders by A. Michal. It should be mentioned that Andrée Bastiani also uses the structure Λ_c of continuous convergence on $\mathcal{L}(E,F)$ in order to define continuity of Df and that Theorem 1.2.2. already appears in her article (cf. [11], pp. 43 f), where a function $f : X \to F$ of class C^1_c is called "différentiable sur X".

1.2.5. Remark. In the definition of functions $f : X \to F$ of class C^1_c the requirement that f be continuous is redundant. The same is true for class C^1_Λ whenever Λ is finer than Λ_c .

Assume that $f : X \to F$ satisfies definition 1.0.0. for $\Lambda = \Lambda_c$, without being continuous a priori. Given $\beta \in \Gamma_F$ and $\varepsilon > 0$, there exists a circled neighbourhood U of 0 in E such that $h \in U$ and $0 \leq t \leq 1$ imply

$$|(Df(x+th)-Df(x))h|_\beta \leq \varepsilon .$$

Then by 1.1.4. (2) we have $|Rf_x(h)|_\beta \leq \varepsilon$ if $h \in U$. This means that Rf_x is continuous at $0 \in X-x$, from which we conclude that $h \longmapsto f(x+h)$ is continuous at 0 ; hence f is continuous, as x is arbitrary. We can further conclude that $Rf : \Omega_X \to F$ is continuous since each of the mappings $(x,h) \longmapsto f(x+h)-f(x)$ and $(x,h) \longmapsto Df(x)h$ is continuous.

1.2.6. Theorem. Assume that $f : X \to F$ is of class C^1_{qb} . Then for every $x \in X$ the remainder $Rf_x : X-x \to F$ satisfies the condition

(FB) $\quad \Theta Rf_x(\mathbb{V}\times\mathfrak{B}) \to 0 \in F$ for every quasi-bounded filter $\mathfrak{B}$ on E .

Here "$\to$" means convergence in the topology of F . We recall that $\mathbb{V}$ denotes the neighbourhood filter of 0 in $\mathbb{R}$ and that a filter $\mathfrak{B}$ on E is said to be quasi-bounded if $\mathbb{V}\cdot\mathfrak{B} \to 0$.

Proof. Given $\beta \in \Gamma_F$, $\varepsilon > 0$ and a quasi-bounded filter $\mathcal{B}$ on E , from the assumption that $Df : X \to \mathcal{L}_{qb}(E,F)$ is continuous we conclude the existence of a circled neighbourhood U of 0 in E and a $B_1 \in \mathcal{B}$ such that $h' \in U$ and $h \in B_1$ imply

$$|(Df(x+h')-Df(x))h|_\beta \leq \varepsilon .$$

We can choose $\delta > 0$ and $B_2 \in \mathcal{B}$ such that $\delta \cdot B_2 \subset U$. Using 1.1.4. (3) we get

$$|\theta Rf_x(t,h)|_\beta \leq \varepsilon \quad \text{if} \quad |t| \leq \delta \quad \text{and} \quad h \in B_1 \cap B_2 .$$

1.2.7. Remark. The condition (FB) has been introduced by A. Frölicher and W. Bucher (cf. [31]) in order to define differentiability of a function between arbitrary c.v.s. (The symbol "FB" appears in Averbukh and Smolyanov's appendix to the Russian translation of Frölicher and Bucher's book, cf. [8]). Our Theorem 1.2.6. shows that Frölicher and Bucher's concept of differentiability fits together with the structure Λ_{qb} of quasi-bounded convergence. However, in order to define "C_1-mappings" from E to F , these authors equip $\mathcal{L}(E,F)$ with the equable convergence structure $\Lambda^{\#}_{qb}$ associated to Λ_{qb} (cf. [31], §§ 9 f). But, according to Theorem 0.5.12., if E and F are l.c.s., $\Lambda^{\#}_{qb}$ coincides with our Π . Thus, in this case $f : E \to F$ is a "C_1-mapping" in the sense of Frölicher and Bucher iff it is of class C^1_Π. As a consequence of the following Theorem 1.2.8. such a function actually satisfies the stronger differentiability condition (HL') of Marinescu and Sebastião e Silva.

1.2.8. Theorem. If $f: X \to F$ is of class C^1_Π, then for every $x \in X$ the remainder $Rf_x: X-x \to F$ satisfies the condition

$$(\text{HL}') \qquad (\forall \beta \in \Gamma_F)(\exists \alpha \in \Gamma_E) \lim_{h\to 0} \frac{|Rf_x(h)|_\beta}{|h|_\alpha} = 0 .$$

Proof. From the assumption that $Df: X \to \mathcal{L}_\Pi(E,F)$ is continuous we deduce that for every $\beta \in \Gamma_F$ there exists an $\alpha \in \Gamma_E$ with the property that for every $\varepsilon > 0$ there is a circled 0-neighbourhood U in E such that

$$|(Df(x+h')-Df(x))h|_\beta \leq \varepsilon|h|_\alpha$$

for every $h' \in U$ and every $h \in E$. The statement (HL') now follows from 1.1 4. (2).

1.2.9. Remarks. (1) The property (HL') had been denoted (MS) in the survey article [6] because it is the condition of differentiability of f at x in the sense of G. Marinescu (cf. [46] and [47]) and coincides with one of the various types of differentiability defined by J. Sebastião e Silva (cf. [53]). In the 1964 article of the present author the condition (HL') had been denoted (F').

(2) Clearly, if $f: X \to F$ is of class C^1_Δ then Rf_x satisfies condition (HL') for every $x \in X$. This had already been proven by G. Marinescu (cf. [47], p. 173). It might be mentioned that the functions of class C^1 in the sense of R. Auwärter-Kuhn (cf. [3]) are of class C^1_Δ.

1.2.10. Remark. The notations (HL') resp. (F') are motivated by the fact that this condition is obtained by an obvious modification of the following stronger remainder condition:

$$\text{(HL)} \qquad (\forall\beta \in \Gamma_F)(\exists\alpha \in \Gamma_E) \lim_{|h|_\alpha \to 0} \frac{|Rf_x(h)|_\beta}{|h|_\alpha} = 0 \,,$$

denoted (F) in [36]. This condition defines differentiation of f at x in the theories of D.H. Hyers (1941), H.R. Fischer (1957, cf. [29]), S. Lang (for general topological v.s. 1962, cf. [43]) and E. Binz (for general c.v.s., 1966, cf. [12]).

1.2.11. Theorem. Assume that $f : X \to F$ is of class C^1_Θ. Then for every $x \in X$ the remainder $Rf_x : X-x \to F$ satisfies the condition

$$\text{(KE')} \qquad (\exists\alpha\in\Gamma_E)(\forall\beta \in \Gamma_F) \lim_{h\to 0} \frac{|Rf_x(h)|_\beta}{|h|_\alpha} = 0 \,.$$

The proof of 1.2.11. is obtained by an obvious slight modification of the proof of 1.2.8.

1.2.12. Remark. The condition (KE'), denoted (M) and called the condition of "Michal differentiability" of f at x in [6], had been suggested by Michal (1940), by Fréchet (1948) and by the present author in his 1964 article [36]; in the latter it is denoted $(F_0^\prime)$.

1.2.13. Remark. As a matter of fact, (KE') is again received by an obvious modification from a stronger condition, namely

$$\text{(KE)} \qquad (\exists \alpha \in \Gamma_E)(\forall \beta \in \Gamma_F) \lim_{|h|_\alpha \to 0} \frac{|Rf_x(h)|_\beta}{|h|_\alpha} = 0 .$$

This condition, denoted (K) in [6], had been introduced and denoted (F_0) by the present author in his 1964 article [36]. According to [8], P. Ver Eecke seems to have given the same definition of differentiability in 1967 (cf. [64]).

1.3. The chain rule

1.3.0. Theorem. Let E , F , G be three l.c.s., let X be open in E and Y open in F . Let Λ , Λ' be convergence structures on $\mathscr{L}(E,F)$, $\mathscr{L}(E,G)$ respectively. Assume that $f : X \to F$ is of class C^1_Λ , that $g : Y \to G$ is of class C^1_k and that $f(X) \subset Y$. Assume further that the mapping

$$(Dg \circ f) \cdot Df : X \to \mathscr{L}_{\Lambda'}(E,G) ,$$

defined by

$$((Dg \circ f) \cdot Df)(x) := Dg(f(x)) \circ Df(x) \quad \text{for each} \quad x \in X ,$$

is continuous. Then $g \circ f : X \to G$ is of class $C^1_{\Lambda'}$, and

$$D(g \circ f) = (Dg \circ f) \cdot Df .$$

Proof. Let $(x,h) \in X \times E$ be arbitrary but fixed. We have to verify that $Dg(f(x))(Df(x)h)$ is the directional derivative of $g \circ f$ at x in the direction h . For $t \in J_{x,h}$, $t \neq 0$, we get, by a short computation, using differentiability of g at $y = f(x)$ in the direction $f(x+th)-f(x)$,

$$t^{-1}((g\circ f)(x+th)-(g\circ f)(x)) = Dg(y)\phi(t)+\Theta Rg_y(t,\phi(t)) ,$$

where $\phi(t) := t^{-1}(f(x+th)-f(x))$. Differentiability of f at x means

$$\lim_{t\to 0} \phi(t) = Df(x)h .$$

Therefore the first term on the right hand of the former equation has limit $Dg(f(x))(Df(x)h)$ if $t \to 0 \in \mathbb{R}$. On the other hand there exist a $\delta > 0$ and a compact set K in F such that $\phi(t) \in K$ if $0 < t \leq \delta$. By the assumption on g , and taking into account that by 1.2.1. the remainder of g at y satisfies the condition (G_k) of "compact differentiability" we have $\lim_{t\to 0} \Theta Rg_y(t,\phi(t)) = 0$. This completes the proof.

1.3.1. <u>Corollary</u>. Let E , F , G , Λ , Λ' be as in 1.3.0. Assume that the composition map

$$\pi : \mathcal{L}_k(F,G)\times\mathcal{L}_\Lambda(E,F) \to \mathcal{L}_{\Lambda'}(E,G)$$

is continuous. Let $f : X \to F$ be of class C^1_Λ , let $g : Y \to G$ be of class C^1_k and assume that $f(X) \subset Y$. Then $g\circ f : X \to G$ is of class $C^1_{\Lambda'}$.

Proof. The mapping $(Dg\circ f)\cdot Df : X \to \mathcal{L}_{\Lambda'}(E,G)$ can be factorized as follows

$$X \xrightarrow{(Dg\circ f,Df)} \mathcal{L}_k(F,G)\times\mathcal{L}_\Lambda(E,F) \xrightarrow{\pi} \mathcal{L}_{\Lambda'}(E,G)$$

and is thus continuous.

1.3.2. <u>Corollary</u>. For each one of the differentiability classes C^1_c , C^1_{qb} , C^1_Π , C^1_Δ , C^1_Θ of functions between open sets in l.c.s. the chain rule holds.

The meaning of this assertion seems clear, e.g. if $f : X \to F$ and $g : Y \to G$ are both of class C^1_c and if $f(X) \subset Y$, then $g \circ f : X \to G$ is of class C^1_c etc. In the case of C^1_c the chain rule has been proven by A. Bastiani (cf. [11]) and in the case of C^1_Π the proof has been carried out by A. Frölicher and W. Bucher (cf. [31]).

Proof. Corollary 1.3.2. follows from 1.3.1. since each of the convergence structures Λ_c , Λ_{qb} , Π , Δ , Θ is finer than Λ_k and because the composition map π is continuous with respect to each of these convergence structures (the same on every occurring function space); cf. the theorems 0.3.5., 0.4.12., 0.5.13., 0.6.9., 0.7.7.respectively.

1.3.3. <u>Corollary</u>. Let E , F , G be three l.c.s., let X be open in E and let Y be open in F . Let $\mathfrak{S}$ be a cover of E by bounded sets. Assume that $f : X \to F$ is of class $C^1_{\mathfrak{S}}$, that $g : Y \to G$ is of class C^1_Π and that $f(X) \subset Y$. Then $g \circ f : X \to G$ is of class $C^1_{\mathfrak{S}}$.

Proof. This follows from 1.3.1. since Π is finer than Λ_k and because by Theorem 0.5.14.

$$\pi : \mathcal{L}_\Pi(F,G) \times \mathcal{L}_{\mathfrak{S}}(E,F) \to \mathcal{L}_{\mathfrak{S}}(E,G)$$

is continuous.

1.3.4. Corollary. Let E be a metrizable l.c.s. and let F, G be arbitrary l.c.s. Let X be open in E and Y open in F. Assume that $f : X \to F$ is of class C^1_s (resp. C^1_k), that $g : Y \to G$ is of class C^1_k and that $f(X) \subset Y$. Then $g \circ f : X \to G$ is of class C^1_s (resp. C^1_k).

Proof. We have to prove that $(Dg \circ f)\cdot Df : X \to \mathcal{L}(E,G)$ is continuous if $\mathcal{L}(E,G)$ is endowed with Λ_s (resp. Λ_k).

(1) Assume that f is of class C^1_s. We have to show that for every convergent sequence $(x_i)_{i\in\mathbb{N}}$ in X with limit x, say, and for every fixed $h \in E$ we have

$$\lim_{i\in\mathbb{N}} Dg(f(x_i))Df(x_i)h = Dg(f(x))Df(x)h .$$

But this is the case since, by the assumption on f, $\lim_{i\in\mathbb{N}} Df(x_i)h = Df(x)h$, and because the sequence $(Dg(f(x_i)))_{i\in\mathbb{N}}$ converges to $Dg(f(x))$ with respect to the topology Λ_k of compact convergence on $\mathcal{L}(F,G)$.

(2) Assume that f is of class C^1_k. If the mapping $(Dg \circ f)\cdot Df : X \to \mathcal{L}_k(E,G)$ were not continuous there would exist a sequence $(x_i)_{i\in\mathbb{N}}$ in X, convergent to some $x \in X$ and a compact set K in E such that the sequence $(Dg(f(x_i))Df(x_i)h)_{i\in\mathbb{N}}$ would not converge to $Dg(f(x))Df(x)h$ uniformly, with respect to h, in K. There is no loss of generality if we assume $Dg(f(x)) \circ Df(x) = 0 \in \mathcal{L}(E,G)$. We could then choose a sequence $(h_i)_{i\in\mathbb{N}}$ in K such that $(Dg(f(x_i))Df(x_i)h_i)_{i\in\mathbb{N}}$ would not converge to 0 in G. Since K is compact and metrizable we can further assume that $(h_i)_{i\in\mathbb{N}}$ is convergent to some $h \in K$. But then we

would have $\lim_{i\in\mathbb{N}} Df(x_i)h_i = Df(x)h$ and therefore

$$\lim_{i\in\mathbb{N}} Dg(f(x_i))Df(x_i)h_i = Dg(f(x))Df(x)h ,$$

thus contradicting the assumptions made above.

1.3.5. Corollary. Let E and F be two l.c.s., let X be open in E and assume that $f : X \to F$ is of class C^1_k . Let $\phi : \Omega \to E$ be of class C^1 (in the "natural sense", cf. 1.0.6.), where Ω is an open set in a finite-dimensional v.s., such that $\phi(\Omega) \subset X$. Then $f \circ \phi : \Omega \to F$ is of class C^1 .

Proof. Corollary 1.3.5. is a direct consequence of 1.3.4.

2. FUNCTIONS OF CLASS C^p

2.0. Preliminary remarks

Let again E and F be two l.c.s. and X an open set in E. We are going to introduce various notions of differentiability of class C^p, $p \in \mathbb{N}$, for functions defined in X with values in F. The definitions will be along the same line as in the case $p = 1$ in the previous chapter. However, a distinction will now be necessary between class $C^p_{\mathfrak{S}}$, $\mathfrak{S}$ a cover of E by bounded sets, and class C^p_{Λ}, Λ a convergence structure finer than Λ_c on the spaces $\mathcal{L}^k(E,F)$. In the case of a function $f : X \to F$ of class $C^p_{\mathfrak{S}}$ the derivatives of f of orders $k \leq p$ will have their values in $\mathcal{H}^k_{\mathfrak{S}}(E,F)$, whereas in the latter case we can choose $\mathcal{L}^k_{\Lambda}(E,F)$. In order to get a common starting-point for all further definitions of differentiability of some class C^p we begin with a very weak notion.

2.1. Weakly p-times differentiable functions

We recall that for each $p \in \mathbb{N}$ the symbol $L^p(E,F)$ denotes the v.s. of all not necessarily continuous p-linear mappings from E^p to F, and that for any $(p,q) \in \mathbb{N}\times\mathbb{N}$ we identify the space $L^p(E,L^q(E,F))$ to $L^{p+q}(E,F)$ by means of the canonical isomorphism $\theta^{p,q}$ of v.s. (c.f. 0.0.1.).

2.1.0. Definition. A function $f : X \to F$ will be called *weakly p-times differentiable* if there exist functions

$$D^k f : X \to L^k(E,F) \; , \; k = 0,1,\ldots,p \; ,$$

such that $D^0 f = f$ and for each $x \in X$, each $h \in E$ and each $k = 0,1,\ldots,p-1$ we have

$$\lim_{t \to 0} t^{-1}(D^k f(x+th) - D^k f(x)) = D^{k+1} f(x)h \; ,$$

where the limit exists with respect to the topology Λ_s of simple convergence in $L^k(E,F)$.

This definition makes sense whenever $p \in \mathbb{N}$, $p \geq 1$. In order to avoid exceptions in the sequel we agree that any function $f : X \to F$ is weakly 0-times differentiable with $D^0 f = f$.

2.1.1. Proposition. If $f : X \to F$ is weakly q-times differentiable ($q \in \mathbb{N}$) then for every $p \in \mathbb{N}$, $p \leq q$, it is weakly p-times differentiable and

$$D^p f : X \to L^p_s(E,F)$$

is weakly (q-p)-times differentiable; furthermore for every $k \in \mathbb{N}$, $k \leq q-p$ we have

$$D^k(D^p f) = D^{p+k} f \; .$$

2.1.2. Proposition. Let $f : X \to F$ be weakly p-times differentiable ($p \in \mathbb{N}$). If

$$D^p f : X \to L_s^p(E,F)$$

is weakly q-times differentiable ($q \in \mathbb{N}$), then f is weakly $(p+q)$-times differentiable and $D^q(D^p f) = D^{p+q} f$.

The Propositions 2.1.1. and 2.1.2. are trivial consequences of the definition 2.1.0.

2.2. Auxiliary formulae

2.2.0. Proposition. Let $f : X \to F$ be weakly p-times differentiable for some $p \in \mathbb{N}$, $p \geq 1$, and assume that $D^p f : X \to L_s^p(E,F)$ is continuous. Let $(x,h) \in X \times E$ be such that $[x,x+h] \subset X$. Then we have

(1) a Taylor's expansion

$$f(x+h) = \sum_{k=0}^{p} \frac{1}{k!} D^k f(x) h^{(k)} + R_p f(x,h) ,$$

where

$$R_p f(x,h) = \frac{1}{(p-1)!} \int_0^1 (1-t)^{p-1} (D^p f(x+th) - D^p f(x)) h^{(p)} dt ;$$

(2) for every $k \in \mathbb{N}$, $k \leq p-1$, and for every choice of $(h_1,\ldots,h_k) \in E^k$

$$(D^k f(x+h) - D^k f(x)) h_1 \ldots h_k = \int_0^1 D^{k+1} f(x+th) h h_1 \ldots h_k dt ;$$

(3) for every $k \in \mathbb{N}$, $k \leq p-1$, for every choice of $(h_1,\ldots,h_k) \in E^k$ and for every $t \in [0,1]$

$$(t^{-1}(D^k f(x+th)-D^k f(x))-D^{k+1}f(x)h)h_1\ldots h_k$$

$$= \int_0^1 (D^{k+1}f(x+\tau th)-D^{k+1}f(x))hh_1\ldots h_k \, d\tau \, .$$

Here $h^{(p)}$ means $(h,\ldots,h) \in E^p$ for $h \in E$.

Proof. The function of a real variable

$$\phi : J_{x,h} = \{t \in \mathbb{R} \mid x+th \in X\} \to F \, ,$$

defined by $\phi(t) := f(x+th)$, is easily seen to be of class C^p in the sense of A.1.2. For $k \leq p$ the derivative $\phi^{(k)} : J_{x,h} \to F$ of ϕ of order k is given by $\phi^{(k)}(t) = D^k f(x+th)h^{(k)}$. Hence A.4.1. yields (1).

Given $k \in \mathbb{N}$, $k \leq p-1$, and fixed $(h_1,\ldots,h_k) \in E^k$ the function $\psi : J_{x,h} \to F$, defined by

$$\psi(t) := D^k f(x+th)h_1\ldots h_k \, ,$$

is of class C^{p-k} in the sense of A.1.2. and has derivative (of order 1) $\psi' : J_{x,h} \to F$ given by

$$\psi'(t) = D^{k+1}f(x+th)hh_1\ldots h_k \, .$$

Hence (2) results from A.3.3.

Finally (3) follows from (2).

2.2.1. <u>Corollary</u>. With the same assumptions as in proposition 2.2.0., for every $\beta \in \Gamma_F$, we have

$$(1)\quad |R_p f(x,h)|_\beta \leq \frac{1}{p!} \sup_{0\leq t\leq 1} |(D^p f(x+th)-D^p f(x))h^{(p)}|_\beta \;;$$

$$(2)\quad |(D^k f(x+h)-D^k f(x))h_1 \ldots h_k|_\beta \leq \sup_{0\leq t\leq 1} |D^{k+1} f(x+th)hh_1\ldots h_k|_\beta \;;$$

$$(3)\quad |(t^{-1}(D^k f(x+th)-D^k f(x))-D^{k+1} f(x)h)h_1\ldots h_k|_\beta \leq$$

$$\leq \sup_{0\leq \tau\leq t} |(D^{k+1} f(x+\tau h)-D^{k+1} f(x))hh_1\ldots h_k|_\beta \;.$$

Proof. We apply A.3.4. to the integrals which occur in 2.2.0.

2.2.2. <u>Corollary</u>. Let $f : X \to F$ be weakly p-times differentiable and assume that $D^p f : X \to L^p_s(E,F)$ is continuous. Then for every $x \in X$ the remainder $h \longmapsto R_p f_x(h) := R_p f(x,h)$ of order p of f at x , as defined by the Taylor's expansion 2.2.0.(1), satisfies

$$(G_s^{(p)})\qquad \lim_{t\to 0} t^{-p} R_p f_x(th) = 0 \qquad \text{for every} \quad h \in E \;.$$

Proof. This is an immediate consequence of 2.2.1.(1) and the assumption on $D^p f$.

2.2.3. <u>Remark</u>. As a matter of fact the assertions of 2.2.1. and 2.2.2. are true if only f is weakly p-times differentiable, without the additional assumption on $D^p f$.

2.3. Functions defined in $\mathbb{R}^m$

Let $f : X \to F$ be weakly p-times differentiable. If we assume, as in 2.2., that $D^p f : X \to L_s^p(E,F)$ is continuous we can not assert that also $D^k f : X \to L_s^k(E,F)$ for $0 \le k < p$ is continuous. If however E has finite dimension this conclusion is correct as we shall see now. In this case we can assume $E = \mathbb{R}^m$ for some $m \in \mathbb{N}$. Then we know that on $L^k(E,F) = \mathcal{L}^k(\mathbb{R}^m,F)$ all the convergence structures introduced in Chapter 0 are identical and $\mathcal{L}^k(\mathbb{R}^m,F)$, endowed with this structure, is an l.c.s. canonically isomorphic to F^{m^k}.

2.3.0. Proposition. Let X be open in $\mathbb{R}^m$ and let F be an l.c.s. Assume that $f : X \to F$ is weakly p-times differentiable and that $D^p f : X \to \mathcal{L}^p(\mathbb{R}^m,F) = F^{m^p}$ is continuous. Then for every $k \in \mathbb{N}$, $0 \le k \le p$,

$$D^k f : X \to \mathcal{L}^k(\mathbb{R}^m,F) = F^{m^k}$$

is continuous.

Proof. By induction on p. The assertion is obviously true if $p = 0$. (In this case the assumption just means that $f : X \to F = \mathcal{L}^0(\mathbb{R}^m,F)$ is continuous). Now assume it to be true for p and let $D^{p+1} f : X \to \mathcal{L}^{p+1}(\mathbb{R},F)$ be continuous. Let $x \in X$ and $\beta \in \Gamma_F$ be fixed. Due to the local compactness of $\mathbb{R}^m$ there exist a circled neighbourhood U of 0 in $\mathbb{R}^m$ and a $\mu > 0$ such that

$$|D^{p+1} f(x+h)h_0 h_1 \cdots h_p|_\beta \le \mu \cdot |h_0| \cdot |h_1| \cdot \ldots \cdot |h_p|$$

for every $h \in U$ and every $(h_0, h_1, \ldots, h_p) \in (\mathbb{R}^m)^{p+1}$. Here $h \longmapsto |h|$ denotes any norm in $\mathbb{R}^m$. If we apply formula 2.2.1.(2) for $k = p$ we get

$$|(D^p f(x+h) - D^p f(x)) h_1 \ldots h_p|_\beta \leq \mu \cdot |h| \cdot |h_1| \cdot \ldots \cdot |h_p| ,$$

which means that $D^p f : X \to \mathscr{L}^p(\mathbb{R}^m, F)$ is continuous. From the hypothesis on p it follows that the assertion of 2.3.0. is true for $p+1$.

2.3.1. Definition. A function $f : X \to F$, where X is open in $\mathbb{R}^m$, is said to be of class C^p if it is weakly p-times differentiable and if $D^p f : X \to \mathscr{L}^p(\mathbb{R}^m, F) = F^{m^p}$ is continuous.

2.3.2. Remarks. (1) A function defined in an open set of $\mathbb{R}^m$ with values in $\mathbb{R}^n$, $(m,n) \in \mathbb{N}\times\mathbb{N}$, is of class C^p, $p \in \mathbb{N}$, iff it is p-times continuously differentiable in the sense of elementary calculus.

(2) A function of one real variable $f : J \to F$, J open in $\mathbb{R}$ and F an arbitrary l.c.s., is of class C^p in the sense of 2.3.1. iff it is of class C^p according to Definition A.1.2. If this is the case then

$$f^{(k)} = D^k f : J \to \mathscr{L}^k(\mathbb{R}, F) = F$$

for every $k \in \mathbb{N}$, $k \leq p$.

2.4. Symmetry of the higher derivatives

2.4.0. Theorem. Let E and F be two l.c.s. and let X be open in E. Assume that $f : X \to F$ is weakly p-times differentiable for some $p \in \mathbb{N}$, $p \geq 1$, and such that $D^p f : X \to L_s^p(E,F)$ is continuous. Then for every $x \in X$ and every $k \in \mathbb{N}$, $0 \leq k \leq p$, the k-linear mapping

$$D^k f(x) : E^k \to F$$

is totally symmetric.

Proof. Let $x \in X$ and $(h_1,\dots,h_k) \in E^k$ be fixed. The set

$$\Omega := \{(t_1,\dots,t_k) \in \mathbb{R}^k \mid x + \sum_{i=1}^{k} t_i h_i \in X\}$$

is open in $\mathbb{R}^k$. For every fixed $u \in F'$, F' the topological dual of F, the real function $\phi : \Omega \to \mathbb{R}$ of k real variables, defined by

$$\phi(t_1,\dots,t_k) := u(f(x + \sum_{i=1}^{k} t_i h_i))$$

is of class C^p. For any choice of the natural numbers $i_1,\dots,i_k \leq k$ we get

$$\partial_{i_1}\partial_{i_2}\dots\partial_{i_k}\phi(0,\dots,0) = u(D^k f(x)h_{i_1}\dots h_{i_k}) ;$$

here the left hand side denotes the partial derivative of ϕ at $(0,\dots,0) \in \Omega \subset \mathbb{R}^k$ of order k, corresponding to the ordered k-tuple $(i_1,i_2,\dots,i_k)$. But from elementary calculus we know that this expression is invariant under any permutation of the finite set $\{1,\ldots,k\}$. Hence, as $u \in F'$ was arbitrary, by the Hahn-Banach theorem it follows that $D^k f(x)$ is totally symmetric.

2.4.1. Remark. From Theorem 2.4.0. it follows that in all the various notions of differentiability of some class C^p to be defined in the sequel the value $D^p f(x)$ of the derivative of order p of a function $f : X \to F$ at some point $x \in X$ will be a totally symmetric p-linear mapping from E^p into F .

2.5. Functions of class $C^p_{\mathfrak{S}}$

2.5.0. Definition. Let $\mathfrak{S}$ denote a collection of bounded sets in E which covers E and let $p \in \mathbb{N}$. A function $f : X \to F$ is said to be differentiable of class $C^p_{\mathfrak{S}}$ if f is weakly p-times differentiable and such that for every $k \in \mathbb{N}$, $0 \leq k \leq p$, the following two conditions are satisfied:

(1) $D^k f(X) \subset \mathcal{H}^k_{\mathfrak{S}}(E,F)$;

(2) $D^k f : X \to \mathcal{H}^k_{\mathfrak{S}}(E,F)$ is continuous.

Clearly f is of class $C^0_{\mathfrak{S}}$ iff it is continuous. For $p = 1$ the definition above yields the notion $C^1_{\mathfrak{S}}$ defined in Section 1.0.

2.5.1. Proposition. Let $f : X \to F$ be of class $C^p_{\mathfrak{S}}$, $p \geq 1$, with respect to some cover $\mathfrak{S}$ of E by bounded sets. Then for each $x \in X$, each $k \in \mathbb{N}$, $k \leq p-1$, and each $S \in \mathfrak{S}$ the limit

$$\lim_{t \to 0} t^{-1}(D^k f(x+th) - D^k f(x)) = D^{k+1} f(x)h$$

exists in the l.c.s. $\mathcal{H}^k_{\mathfrak{S}}(E,F)$, uniformly, with respect to h, in S.

Proof. Given $S \in \mathfrak{S}$, $\beta \in \Gamma_F$ and $\varepsilon > 0$, by formula 2.2.1.(3), using continuity of $D^{k+1} f : X \to \mathcal{L}^{k+1}_{\mathfrak{S}}(E,F)$, we get a $\delta > 0$ such that $t \in \mathbb{R}$, $|t| \leq \delta$ and $h, h_1, \ldots, h_k \in S$ imply

$$|(t^{-1}(D^k f(x+th) - D^k f(x)) - D^{k+1} f(x)h)h_1 \ldots h_k|_\beta \leq \varepsilon .$$

2.5.2. Proposition. Let $f : X \to F$ be of class $C^q_{\mathfrak{S}}$ with respect to some cover $\mathfrak{S}$ of E by bounded sets. Assume that $p \in \mathbb{N}$, $p \leq q$. Then

(1) f is of class $C^p_{\mathfrak{S}}$.

(2) $D^p f : X \to \mathcal{H}^p_{\mathfrak{S}}(E,F)$ is of class $C^{q-p}_{\mathfrak{S}}$.

(3) For each $k \in \mathbb{N}$, $k \leq q-p$ we have

$$D^k(D^p f) = D^{p+k} f : X \to \mathcal{H}^{p+k}_{\mathfrak{S}}(E,F) = \mathcal{H}^k_{\mathfrak{S}}(E, \mathcal{H}^p_{\mathfrak{S}}(E,F)) .$$

Proof. (1) is trivial. To prove (2) and (3) simultaneously, for each $k \leq q-p$ we define $D^k(D^pf) := D^{p+k}f$. We now have to show that for each $x \in X$, each $h \in E$ and each $k \leq q-p-1$ the limit

$$\lim_{t \to 0} t^{-1}(D^k(D^pf)(x+th)-D^k(D^pf)(x)) = D^{k+1}(D^pf)(x)$$

exists in the l.c.s. $L_s^k(E,\mathcal{H}_{\mathfrak{S}}^p(E,F))$. By 2.5.1. this limit even exists in the l.c.s. $\mathcal{H}_{\mathfrak{S}}^k(E,\mathcal{H}_{\mathfrak{S}}^p(E,F))$ whose topology is finer than the topology induced from the former space.

2.5.3. Proposition. Let $\mathfrak{S}$ be a cover of E by bounded sets. Let $f : X \to F$ be of class $C_{\mathfrak{S}}^p$. If $D^pf : X \to \mathcal{H}_{\mathfrak{S}}^p(E,F)$ is of class $C_{\mathfrak{S}}^q$, then f is of class $C_{\mathfrak{S}}^{p+q}$.

Proof. For $k \in \mathbb{N}$, $k \leq q$, we define the continuous functions $D^{p+k}f : X \to \mathcal{H}_{\mathfrak{S}}^{p+k}(E,F)$ by $D^{p+k}f := D^k(D^pf)$. If $k \leq q-1$ we get

$$\lim_{t \to 0} t^{-1}(D^{p+k}f(x+th)-D^{p+k}f(x)) = D^{k+1}(D^pf)(x)h = D^{p+k+1}f(x)h$$

where the limit exists in $\mathcal{H}_{\mathfrak{S}}^{p+k}(E,F) = \mathcal{H}_{\mathfrak{S}}^k(E,\mathcal{H}_{\mathfrak{S}}^p(E,F))$.

Let $\mathfrak{S}$ be a cover of E consisting of bounded sets. For every $p \in \mathbb{N}$ we denote by $C_{\mathfrak{S}}^p(X,F)$ the set of all functions from X to F which are of class $C_{\mathfrak{S}}^p$. The following statement is immediately verified:

2.5.4. <u>Proposition</u>. $C^p_{\mathfrak{S}}(X,F)$ is a linear subspace of the v.s. F^X of all functions from X to F, and for each $k \in \mathbb{N}$, $k \leq p$, the mapping

$$D^k : C^p_{\mathfrak{S}}(X,F) \to C^{p-k}_{\mathfrak{S}}(X,\mathcal{H}^k_{\mathfrak{S}}(E,F))$$

is linear.

If $\mathfrak{S}$ is equal to $\mathfrak{S}_s$, $\mathfrak{S}_k$, $\mathfrak{S}_{pk}$, $\mathfrak{S}_b$, the collection of all finite, compact, precompact, bounded subsets of E respectively, then we write C^p_s, C^p_k, C^p_{pk}, C^p_b respectively for $C^p_{\mathfrak{S}}$. Obviously we always have the inclusions

$$C^p_b(X,F) \subset C^p_{pk}(X,F) \subset C^p_k(X,F) \subset C^p_s(X,F) .$$

2.5.5. <u>Remarks</u>. (1) In the special case that E has <u>finite dimension</u> all these function spaces are identical; they then coincide with the space $C^p(X,F)$ of all functions from X to F which are of class C^p in the sense of 2.3.1.

(2) If E and F are <u>normable</u> l.c.s. then $C^p_b(X,F)$ is the space of all functions from X to F which are Fréchet differentiable of class C^p.

2.6. The differentiability classes C_c^p, C_{qb}^p, C_Π^p, C_Δ^p, C_Θ^p

Let, as in the whole chapter, E and F denote two l.c.s., X an open set in E and p a natural integer. We are going to define a notion of differentiability of class C^p for a function $f : X \to F$ with respect to each of the convergence structures Λ_c, Λ_{qb}, Π, Δ, Θ introduced in Chapter 0 on the occurring spaces $\mathcal{L}^k(E,F)$, $k \in \mathbb{N}$, $k \leq p$, of continuous multilinear mappings $E^k \to F$. Let Λ denote anyone of these convergence structures on the spaces $\mathcal{L}^k(E,F)$, $k \in \mathbb{N}$. We know that for all $(k,q) \in \mathbb{N} \times \mathbb{N}$ one has

$$\mathcal{L}^k(E,\mathcal{L}^q_\Lambda(E,F)) = \mathcal{L}^{q+k}(E,F) \quad \text{if} \quad \Lambda = \Lambda_c , \Lambda_{qb} , \Pi , \Delta ,$$

$$\mathcal{L}^k(E,\mathcal{L}^q_\Theta(E,F)) \subset \mathcal{L}^{q+k}(E,F) ,$$

(cf. 0.6.7., 0.7.6.). In each case we can therefore define C^p_Λ such that the derivatives D^kf of orders $k \leq p$ of a function f of class C^p_Λ have their values in $\mathcal{L}^k(E,F)$.

2.6.0. <u>Definition</u>. A function $f : X \to F$ is said to be <u>differentiable of class</u> C^p_Λ if f is weakly p-times differentiable and such that for each $k \in \mathbb{N}$, $k \leq p$, the following two conditions are satisfied:

(1) $D^kf(X) \subset \mathcal{L}^k(E,F)$;

(2) $D^kf : X \to \mathcal{L}^k_\Lambda(E,F)$ is continuous.

Differentiability of class C^0_Λ just means continuity of the function in question.

For $p = 1$ this definition is consistent with the various concepts of C^1_Λ introduced in Chapter 1.

Clearly C^p_Λ implies C^q_Λ for $q \leq p$.

In the cases where $\Lambda = \Lambda_c$ or Λ_{qb} we write C^p_c resp. C^p_{qb} instead of $C^p_{\Lambda_c}$ resp. $C^p_{\Lambda_{qb}}$.

2.6.1. Proposition. Assume that $f : X \to F$ is weakly p-times differentiable and such that $D^k f(X) \subset \mathcal{L}^k(E,F)$ for every $k \in \mathbb{N}$, $k \leq p$. If

$$D^p f : X \to \mathcal{L}^p_\Lambda(E,F)$$

is continuous, then f is of class C^p_Λ.

Proof. By induction on p. The assertion is trivial for $p = 0$. Assume it to be true for p. Let the hypotheses of 2.6.1. on f be fulfilled for $p+1$. By 2.2.1.(2) for every $\beta \in \Gamma_F$, every $h \in E$ such that $[x,x+h] \subset X$ and every $(h_1,\ldots,h_p) \in E^p$ we have

$$|(D^p f(x+h) - D^p f(x))h_1 \ldots h_p|_\beta \leq \sup_{0 \leq t \leq 1} |D^{p+1} f(x+th)hh_1 \ldots h_p|_\beta .$$

From this inequality we can deduce that $D^p f$ is continuous if $D^{p+1} f$ is. The proof has to be carried out for each of the convergence structures in question separately. If e.g. $\Lambda = \Pi$ then we use the fact that for every $\alpha \in \Gamma_E$ we have

$$|(D^p f(x+h) - D^p f(x)|_{\beta,\alpha} \leq \sup_{0 \leq t \leq 1} |D^{p+1} f(x+th)|_{\beta,\alpha} \cdot |h|_\alpha ;$$

thus $D^p f : X \to \mathcal{L}^p_\Pi(E,F)$ is continuous if $D^{p+1} f : X \to \mathcal{L}^{p+1}_\Pi(E,F)$ is. The remaining cases are left to the reader.

2.6.2. Proposition. Let $f : X \to F$ be of class C^p_Λ . Then for $(x,h) \in X\times E$ and each $k \in \mathbb{N}$, $k \leq p-1$, the limit

$$\lim_{t\to 0} t^{-1}(D^k f(x+th)-D^k f(x)) = D^{k+1}f(x)h$$

exists in the c.v.s. $\mathcal{L}^k_\Lambda(E,F)$.

Proof. We apply formula 2.2.1.(3) and consider the continuity of the function $D^{k+1}f : X \to \mathcal{L}^{k+1}_\Lambda(E,F)$. Again the assertion has to be verified separately for $\Lambda = \Lambda_c, \Lambda_{qb}, \Pi, \Delta, \Theta$.

2.6.3. Remarks. Let E and F be l.c.s., X an open set in E , $f : X \to F$ a mapping and $p \in \mathbb{N}$.

(1) $f \in C^p_c(X,F)$ iff f is "p fois différentiable sur X" according to the definition of A. Bastiani.

(2) $f \in C^p_\Pi(X,F)$ iff f is a "C_p-map" in the sense of A. Frölicher and W. Bucher.

The first of these two statements holds by A. Bastiani's definition (cf. [11], p. 59) which is based on the fact that if $f \in C^p_c(X,F)$ then the derivatives $D^k f$, $0 \leq k \leq p$, as functions from X into the c.v.s. $\mathcal{L}^k_c(E,F)$, are "différentiable sur X" in the sense of A. Bastiani.

In order to verify the second of the statements above, starting from Frölicher-Bucher's definition of "C_p-maps" (cf. [31], 10.1.), and taking into account that by 0.5.12. we have $\Lambda^*_{qb}=\Pi$, it suffices to show that if $f \in C^p_\Pi(X,F)$, then $D^k f$, $0\leq k\leq p$, as functions from X into the c.v.s. $\mathcal{L}^k_\Pi(X,F)$ are "differentiable" in Frölicher-Bucher's sense.

Both cases can easily be settled by means of our formula 2.2.1.(3) which, for each $(x,h) \in X \times E$ such that $[x,x+h] \subset X$, each t such that $0<t\leq 1$, each $(h_1,\ldots,h_k) \in E^k$ and each $\beta \in \Gamma_F$ yields

$$|\theta R(D^k f)_x(t,h)h_1\ldots h_k|_\beta \leq \sup_{0\leq\tau\leq t} |(D^{k+1}f(x+\tau h)-D^{k+1}f(x))hh_1\ldots h_k|_\beta .$$

We omit the details.

2.6.4. Proposition. For every $p \in \mathbb{N}$ the set $C^p_\Lambda(X,F)$ of all functions from X to F which are differentiable of class C^p_Λ is a linear subspace of the v.s. F^X of all functions from X to F, and the mapping

$$D^p : C^p_\Lambda(X,F) \to C^0(X,\mathcal{L}^p_\Lambda(E,F))$$

is linear.

Proof. The verification of this assertion is immediate.

2.7. Relations between the various concepts of C^p_Λ

We are now going to compare the different notions of differentiability of class C^p_Λ which we have established in the previous sections (for fixed but arbitrary $p \in \mathbb{N}$ and variable Λ) for a function $f : X \to F$. We can use the same arguments as in the case $p = 1$. We only have to take into account that $\mathcal{L}^p_{\mathfrak{S}}(E,F)$ is always a subspace of $\mathcal{H}^p_{\mathfrak{S}}(E,F)$ for every cover of E consisting of bounded sets (cf. 0.2.3.) and that these spaces coincide if E is metrizable and $\mathfrak{S}$ contains all compact sets (cf. 0.2.6.) or if E is metrizable and barrelled and $\mathfrak{S}$ is arbitrary (cf. 0.2.7.). We obtain

2.7.0. <u>Theorem</u>. Let E and F be two l.c.s., X an open set in E , $f : X \to F$ a function and $p \in \mathbb{N}$. If Λ is any of the convergence structures Λ_s , Λ_k , Λ_{pk} , Λ_b , Λ_{qb} , Π , Δ or Θ let "C^p_Λ" denote the statement "f is of class C^p_Λ ". Then the statements of Theorem 1.0.2. and of the Corollaries 1.0.3., 1.0.4., 1.0.5., 1.0.6. remain true if we replace everywhere "C^1" by "C^p" .

Especially Table 1 on p.62 remains its validity if we change "C^1" into "C^p" for any $p \in \mathbb{N}$.

2.7.1. <u>Proposition</u>. Let E and F be two l.c.s. and X an open set in E . If $f : X \to F$ is of class C^{p+1}_c for some $p \in \mathbb{N}$, then f is of class C^p_Π .

Proof. We suppose that f is of class C_c^{p+1}, i.e. that f is weakly (p+1)-times differentiable and that $D^{p+1}f : X \to \mathcal{L}_c^{p+1}(E,F)$ is continuous. Then clearly f is weakly p-times differentiable and the only thing we have to prove is that $D^p f : X \to \mathcal{L}_\Pi^p(E,F)$ is continuous. Let $x \in X$ be fixed. Given $\beta \in \Gamma_F$, by 0.3.7., there exist $\alpha \in \Gamma_E$, a circled neighbourhood U of 0 in E and $\mu > 0$ such that $|D^{p+1}f(x+h)|_{\beta,\alpha} \leq \mu$ for all $h \in U$. If we now use formula 2.2.1.(2) we obtain

$$|D^p f(x+h) - D^p f(x)|_{\beta,\alpha} \leq \mu \cdot |h|_\alpha$$

for every $h \in U$. This proves the assertion.

2.7.2. <u>Corollary</u>. For every $p \in \mathbb{N}$ we have the following inclusions of function spaces:

$$C_c^{p+1}(X,F) \subset C_\Pi^p(X,F) \subset C_{qb}^p(X,F) \subset C_b^p(X,F) .$$

2.7.3. <u>Corollary</u>. Let E be a metrizable l.c.s. (resp. a Fréchet space) and F an arbitrary l.c.s. If $f : X \to F$ is of class C_k^{p+1} (resp. of class C_s^{p+1}) for some $p \in \mathbb{N}$, then f is of class C_Π^p.

2.8. Taylor's theorem

2.8.0. Preliminary remarks. Let E and F be l.c.s., X an open set in E and $p \in \mathbb{N}$. From Section 2.2. we deduce that if a function $f : X \to F$ is of class C_s^p, the weakest differentiability class of order p we are considering in this Lecture Note, then for every $(x,h) \in X \times E$ such that $x \in X$ and $x+h \in X$, we have a Taylor's expansion

$$f(x+h) = \sum_{k=0}^{p} \frac{1}{k!} D^k f(x) h^{(k)} + R_p f(x,h) ,$$

where the remainder of order p of f at x,

$$R_p f_x : X - x \to F ,$$

defined by

$$R_p f_x(h) := R_p f(x,h) ,$$

satisfies the Gâteaux condition of order p :

$$(G_s^{(p)}) \qquad \lim_{t \to 0} t^{-p} R_p f_x(th) = 0 \quad \text{for every } h \in E .$$

We are going to deduce properties of remainders of order p for the various differentiability classes of order p which we have introduced, thus generalizing the results of

Section 1.2. With this end in view, the notation being as above, for every $(t,h) \in \mathbb{R}\times E$ such that $x+th \in X$, we put

$$\Theta_p R_p f_x(t,h) := t^{-p} R_p f_x(th) \quad \text{if} \quad t \neq 0 ,$$

$$\Theta_p R_p f_x(0,h) := 0 .$$

We recall that, for any subset S of E, $[S]$ denotes the filter on E, generated by $\{S\}$. As always, $\mathbb{V}$ denotes the filter of 0-neighbourhoods in $\mathbb{R}$.

2.8.1. Theorem. Let E and F be l.c.s., X an open set in E, x a point in X and $p \in \mathbb{N}$. Let $f : X \to F$ be a function, at least of class C_s^p, and denote by $R_p f_x : X-x \to F$ its remainder of order p at x.

(1) Assume $f \in C^p_{\mathfrak{S}}(X,F)$ with respect to some covering $\mathfrak{S}$ of E consisting of bounded sets. Then the following condition $(G^{(p)}_{\mathfrak{S}})$ is satisfied :

$(G^{(p)}_{\mathfrak{S}})$ $\quad \Theta_p R_p f_x(\mathbb{V}\times[S]) \to 0 \in F$ for every $S \in \mathfrak{S}$.

(2) If $f \in C^p_c(X,F)$ the following condition $(MB^{(p)})$ is satisfied:

$(MB^{(p)})$ $\quad \Theta_p R_p f_x(\mathbb{V}\times\mathfrak{X}) \to 0 \in F$ for every convergent filter $\mathfrak{X}$ on E.

(3) If $f \in C^p_{qb}(X,F)$ the following condition $(FB^{(p)})$ is satisfied:

$(FB^{(p)})$ $\quad \Theta_p R_p f_x(\mathbb{V}\times\mathfrak{B}) \to 0 \in F$ for every quasi-bounded filter $\mathfrak{B}$ on E.

(4) If $f \in C^p_{\Pi}(X,F)$ the following condition $(HL'^{(p)})$ is satisfied:

$(HL'^{(p)})$ $\quad (\forall\beta\in\Gamma_F)(\exists\alpha\in\Gamma_E) \quad \lim_{h\to 0} |h|_\alpha^{-p}\cdot|R_p f_x(h)|_\beta = 0$

(5) If $f \in C^p_{\Theta}(X,F)$ **the** following condition $(KE'^{(p)})$ is satisfied:

$(KE'^{(p)})$ $\quad (\exists\alpha\in\Gamma_E)(\forall\beta\in\Gamma_F) \quad \lim_{h\to 0} |h|_\alpha^{-p}\cdot|R_p f_x(h)|_\beta = 0$.

2.8.2. Remarks. (1) For $p = 1$ the statements (1), (2), (3), (4), (5) in 2.8.1. yield 1.2.1., 1.2.3., 1.2.6., 1.2.8., 1.2.10. respectively.

(2) The condition $(G^{(p)}_{\mathfrak{S}})$ means

$(G^{(p)}_{\mathfrak{S}})$ $\quad \lim_{t\to 0} t^{-p} R_p f_x(th) = 0$ uniformly with respect to h in each set $S \in \mathfrak{S}$.

If $\mathfrak{S} = \mathfrak{S}_s , \mathfrak{S}_k , \mathfrak{S}_{pk} , \mathfrak{S}_b$ we write $(G^{(p)}_s)$, $(G^{(p)}_k)$, $(G^{(p)}_{pk})$, $(G^{(p)}_b)$ respectively for $(G^{(p)}_{\mathfrak{S}})$.

(3) The Bastiani condition $(MB^{(p)})$ of order p can be expressed in the form

$$(MB^{(p)}) \qquad \lim_{(t,h)\to(0,h_0)} t^{-p} R_p f_x(th) = 0 \quad \text{for every } h_0 \in E .$$

In [11] the statement 2.8.1.(2) is used as a definition for "application p fois différentiable".

(4) The Marinescu-Sebastião e Silva condition $(HL'^{(p)})$ of order p is derived by a simple modification from the (stronger) Hyers-Lang condition $(HL^{(p)})$ of order p :

$$(HL^{(p)}) \qquad (\forall \beta \in \Gamma_F)(\exists \alpha \in \Gamma_E) \quad \lim_{|h|_\alpha \to 0} |h|_\alpha^{-p} \cdot |R_p f_x(h)|_\beta = 0 .$$

(5) By exactly the same modification as in (4) $(KE'^{(p)})$ is derived from the following stronger condition

$$(KE^{(p)}) \qquad (\exists \alpha \in \Gamma_E)(\forall \beta \in \Gamma_F) \quad \lim_{|h|_\alpha \to 0} |h|_\alpha^{-p} \cdot |R_p f_x(h)|_\beta = 0 .$$

which can also be formulated as follows:

$$(KE^{(p)}) \qquad \lim_{t\to 0} t^{-p} R_p f_x(th) = 0 \quad \text{uniformly with respect to } h \text{ in some 0-neighbourhood.}$$

Proof. The Theorem is obviously true for $p = 0$. If $p \geq 1$ we can apply 2.2.0.; in each occurring case, for every $\beta \in \Gamma_F$ and every $t \in \mathbb{R}$ such that $0 < t \leq 1$, by 2.2.1.(1), we have

$$(*)\quad |t^{-p}R_p f_x(th)|_\beta \leq \frac{1}{p!} \sup_{0 \leq \tau \leq t} |(D^p f(x+\tau h) - D^p f(x))h^{(p)}|_\beta .$$

1) Assume that f is of class $C^p_{\mathfrak{S}}$. Given $\beta \in \Gamma_F$, $\varepsilon > 0$ and $S \in \mathfrak{S}$, using the continuity of $D^p f : X \to \mathcal{H}^p_{\mathfrak{S}}(E,F)$ we can find a circled neighbourhood U of 0 in E such that $h_0 \in U$ and $(h_1,\dots,h_p) \in S^p$ imply

$$|(D^p f(x+h_0) - D^p f(x))h_1 \dots h_p|_\beta \leq p! \cdot \varepsilon .$$

We choose $\delta > 0$ such that $\delta \cdot S \subset U$. If we now apply $(*)$ we see that $|t^{-p}R_p f_x(th)|_\beta \leq \varepsilon$ whenever $0<t\leq\delta$ and $h \in S$.

2) Assume that f is of class C^p_c, i.e. that $D^p f : X \to \mathcal{L}^p_c(E,F)$ is continuous. Given $h_0 \in E$, $\beta \in \Gamma_F$ and $\varepsilon > 0$ there exists a circled neighbourhood U of 0 in E such that $h - h_0 \in U$ and $h_1 \in U$ imply

$$|(D^p f(x+h_1) - D^p f(x))h^{(p)}|_\beta \leq p! \cdot \varepsilon .$$

We choose δ, $0 < \delta \leq 1$, such that $\delta \cdot (h_0 + U) \subset U$. If now $0 < t \leq \delta$ and $h \in h_0 + U$ the inequality $(*)$ yields $|t^{-p}R_p f_x(th)|_\beta \leq \varepsilon$.

3) Assume that f is of class C^p_{qb}, i.e. that $D^p f : X \to \mathcal{L}^p_{qb}(E,F)$ is continuous. Let $\mathfrak{B}$ be a quasi-bounded filter on E. Given $\beta \in \Gamma_F$ and $\varepsilon > 0$ there exist a circled neighbourhood U of 0 in E and $B' \in \mathfrak{B}$ such

that $h_1 \in U$ and $h \in B'$ imply

$$|(D^p f(x+h_1) - D^p f(x))h^{(p)}|_\beta \leq p! \cdot \varepsilon .$$

If we choose δ, $0 < \delta \leq 1$, and $B'' \in \mathcal{B}$ such that $\delta \cdot B'' \subset U$, by the inequality (*) we have $|t^{-p} R_p f_x(th)|_\beta \leq \varepsilon$ for $0 < t \leq \delta$ and $h \in B' \cap B'' \in \mathcal{B}$.

4) If f is of class C^p_Π, i.e. if $D^p f : X \to \mathcal{L}^p_\Pi(E,F)$ is continuous, then given $\beta \in \Gamma_F$ there exists an $\alpha \in \Gamma_E$ with the property that for every $\varepsilon > 0$ there is a circled neighbourhood U of 0 in E such that $h \in U$ implies

$$|D^p f(x+h) - D^p f(x)|_{\beta,\alpha} \leq p! \cdot \varepsilon .$$

If we now apply 2.2.1.(1) we get $|R_p f_x(h)|_\beta \leq \varepsilon \cdot |h|^p_\alpha$ for $h \in U$.

5) The proof of 2.8.1.(5) differs only slightly from that of 2.8.1.(4) and is therefore omitted.

2.8.3. <u>Proposition</u>. Let E and F be l.c.s. and let X be an open set in E. If $f: X \to F$ is of class $C^{p+1}_\Pi(X,F)$ for some $p \in \mathbb{N}$ then, for every $x \in X$, the remainder $R_p f_x : X-x \to F$ of order p of f at x has the property $(HL^{(p)})$ introduced in 2.8.2.(4).

Proof. Let $x \in X$ be fixed. From 2.2.1.(1) and 2.2.1.(2) we deduce that for every $\beta \in \Gamma_F$ and every $h \in E$ such that $[x,x+h] \subset X$ one has

$$|R_p f_x(h)|_\beta \leq \frac{1}{p!} \sup_{0 \leq t \leq 1} |D^{p+1} f(x+th)h^{(p+1)}|_\beta .$$

As $D^{p+1}f : X \to \mathcal{L}_{\Pi}^{p+1}(E,F)$ is assumed to be continuous there exist an $\alpha' \in \Gamma_E$, a circled 0-neighbourhood U in E and $\mu > 0$ such that

$$|D^{p+1}f(x+h)|_{\beta\alpha'} \leq \mu \cdot p!$$

whenever $h \in U$. We then have

$$|R_p f_x(h)|_\beta \leq \mu \cdot |h|_{\alpha'}^{p+1} \quad \text{if} \quad h \in U .$$

We now choose $\alpha \in \Gamma_E$, $\alpha \geq \alpha'$ and such that $h \in U$ whenever $|h|_\alpha \leq 1$. It follows

$$|R_p f_x(h)|_\beta \leq \mu \cdot |h|_\alpha^{p+1} \quad \text{if} \quad |h|_\alpha \leq 1 ,$$

which proves the assertion.

2.9. Functions of class C^∞

2.9.0. Definition. Let E and F be two l.c.s. and X an open set in E . Let Λ denote any of the convergence structures Λ_s , Λ_k , Λ_{pk} , Λ_b , Λ_c , Λ_{qb} , Π , Δ , Θ in the occurring spaces of multilinear mappings. A function $f : X \to F$ is said to be differentiable of class C_Λ^∞ if f is of class C_Λ^p for every $p \in \mathbb{N}$.

Clearly the set

$$C_\Lambda^\infty(X,F) := \bigcap_{p \in \mathbb{N}} C_\Lambda^p(X,F)$$

of all functions of class C_Λ^∞ from X to F is a linear subspace of the v.s. F^X .

It is obvious that between the various notions of C_Λ^∞ (for different convergence structures Λ) we have the same implications as in the case of C^p_Λ for any fixed $p \in \mathbb{N}$ (cf. 2.7.0.). Moreover from 2.7.1. and 2.8.3. we derive the following statements

2.9.1. Theorem. Let E and F be arbitrary l.c.s. and X an open set in E. If $f : X \to F$ is of class C^∞_c then f is of class C^∞_Π and for every $x \in X$ and every $p \in \mathbb{N}$ the remainder $R_p f_x$ of order p of f at x satisfies the condition

$$(HL^{(p)}) \qquad (\forall \beta \in \Gamma_F)(\exists \alpha \in \Gamma_E) \quad \lim_{|h|_\alpha \to 0} |h|_\alpha^{-p} |R_p f_x(h)|_\beta = 0 .$$

2.9.2. Corollary. Let E be a metrizable and F an arbitrary l.c.s. If $f : X \to F$ is of class C^∞_k, then it is of class C^∞_Π.

2.9.3. Corollary. Let E be a Fréchet space and F an arbitrary l.c.s. If $f : X \to F$ is of class C^∞_s, then it is of class C^∞_Π.

2.9.4. Corollary. Let E be a Banach space and F an arbitrary l.c.s. or let E be a Fréchet space and F a normable l.c.s. Then all notions C^∞_Λ which we have introduced, for functions from an open subset X of E into F, coincide.

On Table 2 on the next page we have listed the various notions of differentiability of class C^∞, the relations between them and the remainders of order p which correspond to them for every $p \in \mathbb{N}$.

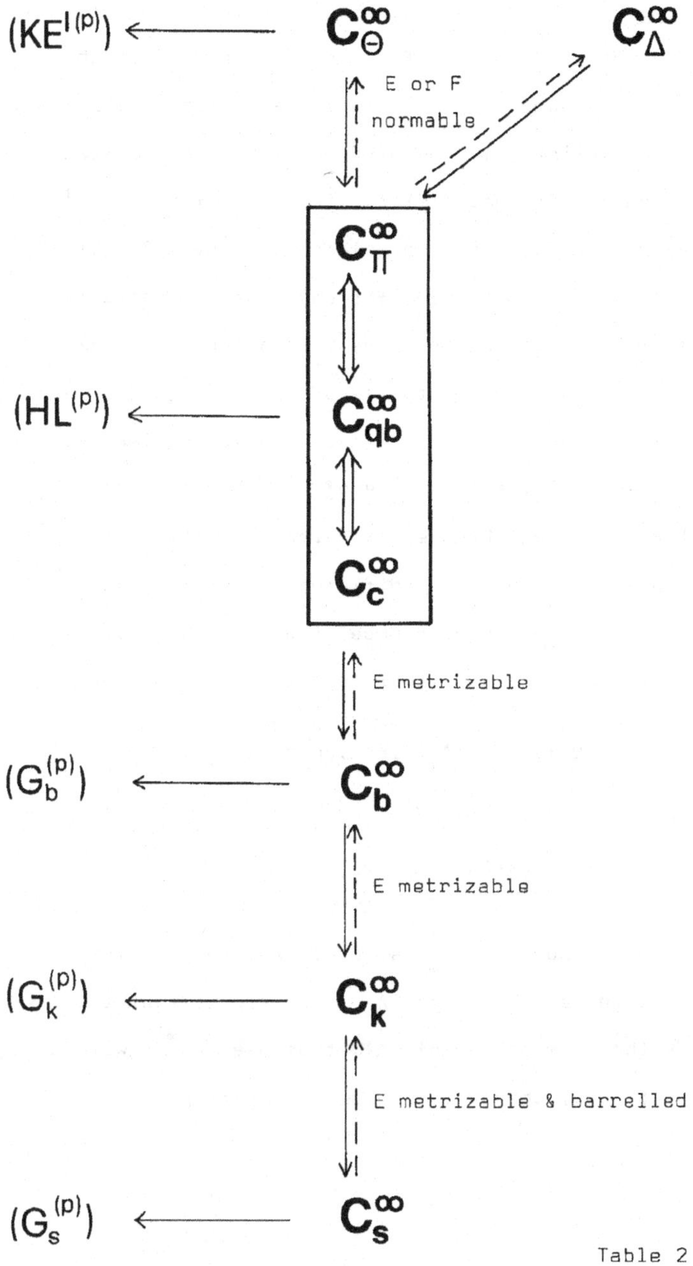
$(KE^{l(p)})$
C^∞_Θ
C^∞_Δ
E or F
normable
C^∞_π
$(HL^{(p)})$
C^∞_{qb}
C^∞_c
E metrizable
$(G_b^{(p)})$
C^∞_b
E metrizable
$(G_k^{(p)})$
C^∞_k
E metrizable & barrelled
$(G_s^{(p)})$
C^∞_s

Table 2

2.9.5. Remarks. (1) In the remarks following Corollary 1.0.3. we have stated that for functions defined in an open subset X of a Banach space E with values in an arbitrary l.c.s. F there exist only two differentiability classes C^1 which may be called Fréchet differentiability of class C^1 and Gâteaux-Lévy differentiability of class C^1. By 2.7.0. the same is true if we consider C^p for any $p \in \mathbb{N}$ instead of C^1. However from 2.7.1. we conclude that a function $f:X \to F$ which is Gâteaux-Lévy differentiable of class C^{p+1} for some $p \in \mathbb{N}$ actually is even Fréchet differentiable of class C^p at least. Consequently, in the case of C^∞ the difference between Gâteaux-Lévy and Fréchet differentiabilities disappears and we may speak of a function "differentiable of class C^∞" without reference to any convergence structure Λ. By definition we have $f \in C^\infty(X,F)$ iff there exist continuous functions

$$D^p f : X \to \mathcal{L}_s^p(E,F) \ , \ p = 0,1,2,\ldots$$

such that $D^{p+1}f$ is, for each $p \in \mathbb{N}$, the Gâteaux-Lévy (i.e. the directional) derivative of $D^p f$.

(2) If F is a normable l.c.s. or if we are only interested in convergence structures Λ which are coarser than Π, we may apply the terminology of (1), thus use "C^∞" without a further attribute, whenever E is a Fréchet space.

2.10. Higher order chain rule

In this section we shall generalize the chain rule 1.3.0. to functions of class C^p , $p \in \mathbb{N}$. However, for the sake of simplicity, we shall restrict ourselves to the differentiability classes which are stronger or equal to C^p_c . The main Theorem states that in each of these classes, and for every $p \in \mathbb{N}$, a p-th order chain rule holds.

2.10.0. Theorem. Let E , F and G be l.c.s., let X be open in E , let Y be open in F and let p be a natural number. Let Λ denote any one of the convergence structures Λ_c , Λ_{qb} , Π , Δ or Θ , the same on each of the occurring spaces $\mathcal{L}^q(E,F)$, $\mathcal{L}^q(F,G)$, $\mathcal{L}^q(E,G)$, $q \in \mathbb{N}$, $q \leq p$.

Assume that $f : X \to F$ and $g : Y \to G$ are both of class C^p_Λ and that $f(X) \subset Y$. Then $g \circ f : X \to G$ is of class C^p_Λ , and for every $q \in \mathbb{N}$, $q \leq p$, its derivative $D^q(g \circ f)$ of order q is a finite sum of continuous functions of the form

$$(*) \qquad (D^r g \circ f) \cdot D^{i_1} f \cdot D^{i_2} f \cdot \ldots \cdot D^{i_r} f : X \to \mathcal{L}^q_\Lambda(E,G) ,$$

where $r \in \mathbb{N}$, $1 \leq r \leq q$, and where $(i_1, i_2, \ldots, i_r)$ is an r-tuple of strictly positive integers with sum q .

Proof. The interpretation of the expression (*) should be clear. It is the composition of two mappings

$$X \to \mathcal{L}^r_\Lambda(F,G) \times \mathcal{L}^{i_1}_\Lambda(E,F) \times \ldots \times \mathcal{L}^{i_r}_\Lambda(E,F) \to \mathcal{L}^q_\Lambda(E,G),$$

the first being the function $(D^r g \circ f, D^{i_1} f, \ldots, D^{i_r} f)$ which is continuous by the assumptions on f and g , whereas the second, the canonical $(r+1)$-linear mapping π , is continuous due to 0.3.5., 0.4.12., 0.5.13., 0.6.9. or 0.7.7. Hence $(*)$ is continuous.

The case $p = 0$ is trivial. Assume now $p \geq 1$. We proceed by induction on p . In case $p = 1$, by 1.3.2., $g\circ f$ is of class C^1_Λ and $D(g\circ f) = (Dg\circ f)\cdot Df$ has the required form.

Let us assume that the assertion of the Theorem is true for some p and that f and g are both of class C^{p+1}_Λ . All we have to do is to show that for each (r,q) such that $1 \leq r \leq q \leq p$ the function $(*)$ has a Gâteaux-Lévy derivative which is a sum of expressions of the form $(*)$ with q replaced by $q+1$. Indeed, by a computation similar to that in the proof of 1.3.0., the Gâteaux-Lévy derivative of $(*)$ turns out to be

$$(D^{r+1}g\circ f)\cdot Df\cdot D^{i_1}f\cdot\ldots\cdot D^{i_r}f + (D^r g\circ f)\cdot D^{i_1+1}f\cdot D^{i_2}f\cdot\ldots\cdot D^{i_r}f +$$
$$+ (D^r g\circ f)\cdot D^{i_1}f\cdot D^{i_2+1}f\cdot\ldots\cdot D^{i_r}f + \ldots + (D^r g\circ f)\cdot D^{i_1}f\cdot D^{i_2}f\cdot\ldots\cdot D^{i_r+1}f.$$

The verification of this last assertion essentially uses the fact that $D^r g : Y \to \mathcal{L}^r_s(F,G)$ is of class C^1_Λ , thus at each $y \in Y$ has a remainder which satisfies (MB) (whereas (G_k) would do). The details are left to the reader.

2.10.1. <u>Corollary</u>. Let $f : X \to F$ and $g : Y \to G$ be both of class C^∞_c (resp. C^∞_Θ , resp. C^∞_Δ) and such that $f(X) \subset Y$. Then $g\circ f : X \to G$ is of class C^∞_c (resp. C^∞_Θ , resp. C^∞_Δ).

2.10.2. Corollary. Let E and F be Fréchet spaces, let G be an arbitrary l.c.s., let X , Y be open subsets of E respectively. If $f : X \to F$ and $g : Y \to G$ are both of class C_s^∞ and such that $g(X) \subset Y$, then $g \circ f : X \to G$ is of class C_s^∞ .

2.10.3. Corollary. Let E and F be Banach spaces and G an arbitrary l.c.s., or let E and F be Fréchet spaces and G a normable l.c.s., let X,Y be open subsets of E,F respectively. Assume that $f : X \to F$ and $g : Y \to G$ are both of class C^∞ (in the sense of 2.9.5.) and such that $f(X) \subset Y$. Then $g \circ f : X \to G$ is of class C^∞ .

APPENDIX: FUNCTIONS OF A REAL VARIABLE

This is a short survey on some simple results in the differential and integral calculus of functions of a real variable with values in an l.c.s. We do not give proofs here. The reader is referred to the standard literature, e.g. [15] and [20]. It might be emphasized that we are not interested here in the strongest possible form of propositions and theorems. For example it is well known that the assertion of A.3.4. is still true if the function f is only assumed to be continuous in $[a,b]$ and differentiable in the interior of $[a,b]$; continuity of f' is redundant.

The symbol J will always denote an open subset of $\mathbb{R}$, in some statements it will be required to be connected, i.e. an open interval. By F we shall denote a (Hausdorff) locally convex space (l.c.s.) and by $\tilde{F}$ the completion of F .

A.1. Derivatives

A.1.1. Definition. A function $f : J \to F$ is said to be of class C^1 if there exists a continuous function $f' : J \to F$, the derivative of f , such that for every $t \in J$

$$\lim_{\tau\to 0} \frac{1}{\tau}(f(t+\tau)-f(t)) = f'(t) .$$

A.1.2. Definition. Let $f : J \to F$ be a function. By recursion we define f to be of class C^r , $r \in \mathbb{N}$, as follows:

(1) f is of class C^0 if it is continuous.

(2) f is of class C^r, $r \geq 1$, if f is of class C^1 and if f' is of class C^{r-1}.

If f is of class C^r then $f^{(r)} : J \to F$, the <u>derivative of</u> f <u>of order</u> r, is defined inductively as follows:

(1') $f^{(0)} = f$

(2') $f^{(r)} = (f^{(r-1)})'$ if $r \geq 1$.

The function $f : J \to F$ is said to be <u>of class</u> C^∞, if it is of class C^r for every $r \in \mathbb{N}$.

A.1.3. <u>Proposition</u>. Let $f : J \to F$ be a function.

(1) If f is of class C^r, and if $k \leq r$, then f is of class C^k, $f^{(k)}$ is of class C^{r-k} and $(f^{(k)})^{(r-k)} = f^{(r)}$.

(2) If f is of class C^k, and if $f^{(k)}$ is of class C^ℓ $(k \in \mathbb{N}, \ell \in \mathbb{N})$, then f is of class $C^{k+\ell}$ and $f^{(k+\ell)} = (f^{(k)})^{(\ell)}$.

A.1.4. <u>Proposition</u>. The set $C^r(J,F)$ of all functions of class C^r from J to F is a subspace of the v.s. F^J, and if $0 \leq k \leq r$, the assignment $f \longmapsto f^{(k)}$ is a linear mapping from $C^r(J,F)$ into $C^{r-k}(J,F)$.

A.1.5. <u>Proposition</u>. Let $(F_i)_{i \in I}$ be a family of l.c.s. A function

$$f = (f_i)_{i \in I} : J \to \prod_{i \in I} F_i$$

is of class C^r $(r \in \mathbb{N})$ iff $f_i : J \to F$ is of class C^r for every $i \in I$. If this condition is fulfilled, then $f^{(r)} = (f_i^{(r)})_{i \in I}$.

A.1.6. <u>Corollary</u>. $C^r(J, \prod_{i\in I} F_i) = \prod_{i \in I} C^r(J,F_i)$.

A.1.7. <u>Proposition</u>. Let F_1, F_2 and F be l.c.s. Assume that a continuous bilinear mapping $(y_1,y_2) \longmapsto y_1 \cdot y_2$ from $F_1 \times F_2$ into F is given. If $f_1 : J \to F_1$ and $f_2 : J \to F_2$ are of class C^r, then $f_1 \cdot f_2 : J \to F$, defined by $(f_1 \cdot f_2)(t) = f_1(t) \cdot f_2(t)$, is of class C^r and Leibniz's formula holds:

$$(f_1 \cdot f_2)^{(r)} = \sum_{k=0}^{r} \binom{r}{k} f_1^{(k)} \cdot f_2^{(r-k)} .$$

For every $r \in \mathbb{N}$ and for $r = \infty$ the space $C^r(J,\mathbb{R})$ of all (real valued) functions of class C^r of a real variable in J, together with the usual addition and multiplication of functions, is a commutative ring with unit and an $\mathbb{R}$-algebra. If we apply A.1.7. to scalar multiplication $\mathbb{R} \times F \to F$ we get

A.1.8. <u>Corollary</u>. For every $r \in \mathbb{N}$ and for $r = \infty$ the space $C^r(J,F)$ is a unitary $C^r(J,\mathbb{R})$-module.

The assertion of A.1.7. extends to products of any finite number m of functions. A very special case is $m = 1$:

A.1.9. <u>Proposition</u>. Let F and G denote two l.c.s. Let $u : F \to G$ be a continuous linear mapping. If $f : J \to F$ is of class C^r, then $u \circ f : J \to G$ is of class C^r and $(u \circ f)^{(r)} = u \circ f^{(r)}$.

A.1.10. Proposition (Chain rule). Let $\phi : J \to \mathbb{R}$ and $f : J_1 \to F$ be of class C^r. Assume that $\phi(J) \subset J_1$, where J and J_1 denote open subsets of $\mathbb{R}$. Then $f \circ \phi : J \to F$ is of class C^r and $(f \circ \phi)' = \phi' \cdot (f' \circ \phi)$.

A.2. Integrals

Let $[a,b] := \{t \in \mathbb{R} \mid a \leq t \leq b\}$ be a compact interval on $\mathbb{R}$, F an arbitrary l.c.s. and $f : [a,b] \to F$ a function. In order to define the Riemann integral of f, given any subdivision

$$\mathcal{E}: a = t_0 < t_1 < \ldots < t_n = b$$

of $[a,b]$, where n may depend on $\mathcal{E}$, we consider the set

$$S_{\mathcal{E}} := \sum_{i=1}^{n} (t_i - t_{i-1}) \cdot f([t_{i-1}, t_i]) \subset F .$$

For every $\delta > 0$ we put

$$S(\delta) := \bigcup_{|\mathcal{E}| \leq \delta} S_{\mathcal{E}} \subset F ,$$

where the union is taken over all subdivisions $\mathcal{E}$ of $[a,b]$ with the property

$$|\mathcal{E}| := \max_{1 \leq i \leq n} |t_i - t_{i-1}| \leq \delta .$$

It is obvious that the family $(S(\delta))_{\delta > 0}$ of sets is a filter-basis on F. The filter on F, generated by $(S(\delta))_{\delta>0}$, will be called the Riemann filter of the function f.

A.2.1. Proposition and Definition. Let $f : [a,b] \to F$ be a continuous function. Then the Riemann filter of f is a Cauchy filter on F and thus has a well-defined limit in the completion $\tilde{F}$ of F .

This limit will be called the <u>Riemann integral of</u> f (shortly: the integral of f) and it will be denoted by either of the symbols

$$\int_a^b f(t)dt \quad \text{or} \quad \int_a^b f .$$

We also define

$$\int_b^a f(t)dt = \int_b^a f := -\int_a^b f .$$

If F is a complete l.c.s. then the integral of f is an element of F . As a matter of fact this is already the case when F is sequentially complete, since the Riemann filter of f has a countable basis.

We shall denote by $C^0([a,b],F)$ the v.s. of all continuous functions from $[a,b]$ to F .

A.2.2. <u>Proposition</u>. The assignment $f \longmapsto \int_a^b f$ defines a linear mapping from $C^0([a,b],F)$ into $\tilde{F}$.

A.2.3. <u>Proposition</u>. Let a , b , c be real numbers such that $a < b < c$, and assume that $f \in C^0([a,c],F)$. Then

$$\int_a^c f = \int_a^b f + \int_b^c f .$$

A.2.4. <u>Proposition</u>. Assume $\phi \in C^0([a,b],\mathbb{R})$ and $y_o \in F$. Then

$$\int_a^b (\phi(t)\cdot y_o)dt = (\int_a^b \phi)\cdot y_o .$$

A.2.5. Proposition. Assume $f \in C^0([a,b],F)$ and $u \in \mathcal{L}(F,G)$, where G is an l.c.s. with completion $\tilde{G}$. Let $\tilde{u} : \tilde{F} \to \tilde{G}$ denote the continuous extension of u . Then

$$\int_a^b (u \circ f) = \tilde{u}\left(\int_a^b f\right) .$$

A.2.6. Theorem (Mean value theorem). Let $f : [a,b] \to F$ be continuous. Assume that M is a convex subset of F such that $f([a,b]) \subset M$. Then

$$\int_a^b f \in (b-a)\cdot\tilde{M} .$$

Here $\tilde{M}$ denotes the completion of M , i.e. the closure of M in $\tilde{F}$.

A.2.7. Corollary. Let $f : [a,b] \to F$ be continuous. For every continuous semi-norm β on F we have

$$\left|\int_a^b f\right|_\beta \leq \int_a^b |f(t)|_\beta dt \leq (b-a)\cdot \sup_{a \leq t \leq b} |f(t)|_\beta .$$

Here, for the sake of convenience, we have denoted the continuous extension of β to $\tilde{F}$ by the same symbol β .

In order to give a more elegant form to A.2.7. we introduce the topology of uniform convergence on the v.s. $C^0([a,b],F)$. This is the l.c. topology defined by the semi-norms

$$f \longmapsto |f|_\beta := \sup_{a \leq t \leq b} |f(t)|_\beta , \ \beta \in \Gamma_F .$$

The l.c.s. $C^0([a,b],F)$ is complete iff F is complete. Now the formula of A.2.7. reads

$$\left|\int_a^b f\right|_\beta \leq (b-a)|f|_\beta ,$$

and, together with A.2.2., yields

A.2.8. Corollary. Let $[a,b]$ be a compact interval on $\mathbb{R}$ and let F be an l.c.s. The assignment $f \longmapsto \int_a^b f$ is a continuous linear mapping from the l.c.s. $C^0([a,b],F)$ of all continuous functions $f : [a,b] \to F$, endowed with uniform convergence, into the completion $\tilde{F}$ of F.

A.3. Relationship between integral and derivative

A.3.1. Theorem. Let J be an open interval on $\mathbb{R}$, F an l.c.s. and $f \in C^0(J,F)$. For every $a \in J$ the function $\int_a f : J \to \tilde{F}$, defined by

$$\left(\int_a f\right)(t) := \int_a^t f , \quad \text{for every } t \in J ,$$

is of class C^1 and

$$\left(\int_a f\right)' = f .$$

A.3.2. Corollary. If $f \in C^r(J,F)$, $r \in \mathbb{N}$, then $\int_a f \in C^{r+1}(J,\tilde{F})$ and

$$\left(\int_a f\right)^{(k)} = f^{(k-1)} \qquad (1 \leq k \leq r+1) .$$

A.3.3. Theorem. Let J be an open interval on $\mathbb{R}$, F an l.c.s. and $f \in C^1(J,F)$. Then for every $(a,b) \in J\times J$

$$f(b)-f(a) = \int^b f' .$$

Remark. The integral on the right hand side of the last formula is an element of F whether or not F is sequentially complete.

A.3.4. Corollary. With the hypotheses of A.3.3., for every continuous semi-norm β on F, one has

$$|f(b)-f(a)|_\beta \leq \int_a^b |f'(t)|_\beta \cdot |dt| \leq |b-a| \cdot \sup_{t\in I} |f'(t)|_\beta \ .$$

Here $I := [a,b]$ if $a \leq b$ and $I := [b,a]$ if $b \leq a$.

A.3.5. Corollary (Fundamental theorem of calculus). Let J be an open interval on $\mathbb{R}$ and F an l.c.s. If $f \in C^1(J,F)$ is such that $f' = 0$, then there exists $y_0 \in F$ such that $f(t) = y_0$ for every $t \in J$.

A.4. Taylor's theorem

A.4.1. Theorem. Let J be an open interval in $\mathbb{R}$, let F be an l.c.s. and assume that $f \in C^n(J,F)$ for some $n \in \mathbb{N}$, $n \geq 1$. Then for every $s \in J$ and every $t \in \mathbb{R}$ such that $s+t \in J$ we have a Taylor's expansion

$$f(s+t) = \sum_{k=0}^{n} \frac{t^k}{k!} f^{(k)}(s) + R_n f(s,t) \ ,$$

where the remainder $R_n f$ of order n has the following representation

$$R_n f(s,t) = \frac{t^n}{(n-1)!} \int_0^1 (1-\tau)^{n-1} (f^{(n)}(s+\tau t) - f^{(n)}(s)) d\tau \ .$$

A.4.2. <u>Corollary</u>. For each $\beta \in \Gamma_F$ the remainder $R_n f$ of order n in the Taylor's expansion of f at $s \in J$ satisfies the inequality:

$$|R_n f(s,t)|_\beta \le \frac{|t|^n}{n!} \sup_{0 \le \tau \le 1} |f^{(n)}(s+\tau t) - f^n(t)|_\beta \, .$$

A.4.3. <u>Corollary</u>. For each fixed $s \in J$ the remainder $R_n f$ of order n in the Taylor's expansion of f at s satisfies

$$\lim_{t \to 0} t^{-n} R_n f(s,t) = 0 \, .$$

BIBLIOGRAPHY

[1] C. Apostol: A theorem of existence and uniqueness for equations with differentials in locally convex spaces. An. Univ. Bucureşti, Ser. Şti. Natur. Mat.-Mec. 13 (1964) no.2, 45-53. MR 32 # 2940.

[2] C. Apostol: On a theorem of existence and uniqueness. An. Univ. Bucureşti, Ser. Şti. Natur. Mat.-Mec. 15 (1966) no.1, 137-141. MR 36 # 1992.

[3] R. Auwärter-Kuhn: Differentialrechnung in lokalkonvexen und Marinescu-Räumen. Diss. Univ. Zürich 1972.

[4] V.I. Averbukh and O.G. Smolyanov: Differentiation in linear topological spaces. Soviet Math. Dokl. 8 (1967) 444-448. MR 35 # 5934.

[5] V.I. Averbukh and O.G. Smolyanov: The theory of differentiation in linear topological spaces. Russian Math. Surveys 22 (1967) no.6, 201-258. MR 36 # 6933.

[6] V.I. Averbukh and O.G. Smolyanov: The various definitions of the derivative in linear topological spaces. Russian Math. Surveys 23 (1968) no.4, 67-113. MR 39 # 7424.

[7] V.I. Averbukh: Higher order derivatives in linear topological spaces. Vestnik Moskov. Univ. Ser. I. Mat. Meh. 25 (1970) no.1, 29-32. MR 43 # 3802.

[8] V.I. Averbukh and O.G. Smolyanov: Appendix to the Russian translation of [31]. MIR, Moscow 1970 (Russian).

[9] M. Balanzat: Theory of the differential in the sense of Hadamard-Fréchet for mappings between topological vector spaces. I. Rev. Un. Mat. Argentina 20 (1962) 155-187. MR 27 # 2840. II. Math. Notae 19 (1964) 43-62. MR 30 # 426.

[10] A. Bastiani: Différentiabilité dans les espaces localement convexes. Distructures. Thèse Doct. Sci. Math. Fac. Sci. Univ. Paris 1962.

[11] A. Bastiani: Applications différentiables et variétés différentiables de dimension infinie. J. Anal. Math. 13 (1964) 1-114. MR 31 # 1540.

[12] E. Binz: Ein Differenzierbarkeitsbegriff in limitierten Vektorräumen. Comment. Math. Helv. 41 (1966) 137-156. MR 34 # 8144.

[13] E. Binz und H.H. Keller: Funktionenräume in der Kategorie der Limesräume. Ann. Acad. Sci. Fenn. AI, 383 (1966) 21pp. MR 34 # 6706.

[14] E. Binz und W. Meier-Solfrian: Zur Differentialrechnung in limitierten Vektorräumen. Comment. Math. Helv. 42 (1967) 285-296. MR 37 # 756.

[15] N. Bourbaki: Fonctions d'une variable réelle. Chap.I,II. ASI 1074. Hermann, Paris 1958.

[16] J.P. Bourguignon: Calcul différentiel généralisé. Séminaire Choquet 1969/70; Initiation à l'Analyse. Fasc.2, Exp.11, 18pp. Secrétariat math., Paris 1970. MR 44 # 2034.

[17] W. Bucher: Différentiabilité de la composition et complétitude de certains espaces fonctionnels. Comment. Math. Helv. 43 (1968) 256-288. MR 37 # 4617.

[18] J.F. Colombeau: Différentiation et bornologie. Thèse, Université de Bordeaux 1973.

[19] J.F. Colombeau: Inversion d'une application différentiable entre espaces bornologiques. C.R. Acad. Sci. Paris, Sér.A 270 (1970) 1692-1694. MR 43 # 3803.

[20] J. Dieudonné: Foundations of Modern Analysis. Acad. Press, New York, 1960.

[21] M.D. Donsker and J.L. Lions: Fréchet-Volterra variational equations, boundary value problems and function space integrals. Acta Math. 108 (1962) 147-228. MR 27 # 170.

[22] E. Dubinsky: Differential calculus and differential equations in Montel spaces. Thesis, Univ. of Michigan, 1962.

[23] E. Dubinsky: Fixed points in non-normed spaces. Ann. Acad. Sci. Fenn. AI, 331 (1963) 6pp. MR 27 # 1801.

[24] E. Dubinsky: Differential equations and differential calculus in Montel spaces. Trans. Am. Math. Soc. 110 (1964) 1-21. MR 29 # 494.

[25] P.L. Falb and M.Q. Jacobs: On differentials in locally convex spaces. J. Differential Equations 4 (1968) 444-459. MR 37 # 5696.

[26] P.L. Falb and M.Q. Jacobs: Correction: "On differentials...". J. Differential Equations 6 (1969) 395-396. MR 41 # 820.

[27] S. Fernandez Long de Foglio: La différentielle au sens d'Hadamard dans les espaces L vectoriels. Portugal. Math. 19 (1960) 165-184. MR 23 # A 2070.

[28] R. Féron: Dérivation sur les espaces affines localement convexes sur $\mathbb{R}$ ou $\mathbb{C}$. Portugal. Math. 27 (1968) 43-53. MR 41 # 7431.

[29] H.R. Fischer: Differentialkalkül für nicht-metrische Strukturen. I. Ann. Acad. Sci. Fenn. AI, 247 (1957) 15pp. MR 19 - 869. II. Archiv der Math. 8 (1957) 428-443. MR 20 # 6669.

[30] H.R. Fischer: Limesräume. Math. Ann. 137 (1959) 269-303. MR 22 # 225.

[31] A. Frölicher and W. Bucher: Calculus in Vector Spaces without Norm. Lecture Notes in Math. 30, Springer, Berlin, 1966. MR 35 # 4723.

[32] W. Gähler: Eine Verallgemeinerung der Differenzierbarkeit I. Math. Nachr. 38 (1968) 217-256. MR 38 # 6003.

[33] J. Gil de Lamadrid: Topology of mappings in locally convex topological vector spaces, their differentiation and integration and application to gradient mappings. Thesis, Univ. of Michigan, 1955.

[34] J. Gil de Lamadrid: Topology of mapping and differentiation processes. Illinois J. Math. 3 (1959) 408-420. MR 21 # 5918.

[35] H. Hogbe-Nlend: Théorie des bornologies et applications. Lecture Notes in Math. 213, Springer, Berlin, 1971.

[36] H.H. Keller: Differenzierbarkeit in topologischen Vektorräumen. Comment. Math. Helv. 38 (1964) 308-320. MR 29 # 3858.

[37] H.H. Keller: Räume stetiger multilinearer Abbildungen als Limesräume. Math. Ann. 159 (1965) 259-270. MR 33 # 1695.

[38] H.H. Keller: Ueber Probleme, die bei einer Differentialrechnung in topologischen Vektorräumen auftreten. Festband z. 70.Geburtstag v. Rolf Nevanlinna, Springer Berlin 1966, 49-57. MR 36 # 6934.

[39] H.H. Keller: Limit vector spaces of Marinescu type. Rev. Roumaine Math. Pures Appl. 13 (1968) 1107-1112. MR 38 # 4944.

[40] J. Kijowski and W. Szczyrba: On differentiability in an important class of locally convex spaces. Studia Math. 30 (1968) 247-257. MR 38 # 1524.

[41] J. Kijowski and J. Komorowski: A differentiable structure in the set of all bundle sections over compact subsets. Studia Math. 32 (1969) 191-207. MR 39 # 3528.

[42] J. Kijowski: Existence of differentiable structure in the set of submanifolds. Studia Math. 33 (1969) 93-108. MR 39 # 6360.

[43] S. Lang: Introduction to Differentiable Manifolds. Interscience Publ., Wiley, New York, 1962.

[44] J.A. Leslie: On a differential structure for the group of diffeomorphisms. Topology 6 (1967) 263-271. MR 35 # 1041.

[45] J.A. Leslie: Some Frobenius theorems in global analysis. J. Diff. Geom. 2 (1968) 279-297. MR 40 # 4977.

[46] G. Marinescu: Différentielles de Gâteaux et Fréchet dans les espaces localement convexes. Bull. Math. Soc. Sci. Math. Phys. RP Roumaine 1 (1957) 77-86. MR 20 # 1188.

[47] G. Marinescu: Espaces vectoriels pseudotopologiques et théorie des distributions. VEB Deutsch. Verlag d. Wissensch., Berlin, 1963. MR 29 # 3878.

[48] D. Meeùs: Le calcul différentiel dans les espaces à bornés convexes. Thèse, Louvain, 1970.

[49] D. Meeùs: Sur la dérivée d'une fonction entre parties d'espaces localement convexes. C.R. Acad. Sc. Paris, Sér.A, 271 (1970) 1250-1253. MR 42 # 8276.

[50] D. Meeùs: Fonctions implicites, non locales, dans les espaces localement convexes. C.R. Acad. Sc. Paris, Sér.A., 272 (1971) 724-726. MR 45 # 9223.

[51] F. und R. Nevanlinna: Absolute Analysis. Springer, Berlin, 1959.

[52] J. Sebastião e Silva: Le calcul différentiel et intégral dans les espaces localement convexes, réels ou complexes. I., II. Atti Accad. Lincei Rend. Cl. Sci. Fis. Mat. Natur. 20 (1956) 743-750; ibid. 21 (1956) 40-46. MR 19 - 561.

[53] J. Sebastião e Silva: Conceitos de função differenciavel em espaços localmente convexos. Publ. Centro Estudes Matem. Lisboa Inst. Alta Cultura, Lisboa, 1957. MR 21 # 283.

[54] J. Sebastião e Silva: Les espaces à bornés et la notion de fonction différentiable. Coll. Analyse-fonct., Louvain, 1960, 57-61. MR 25 # 3354.

[55] U. Seip: Kompakt erzeugte Vektorräume und Analysis. Lecture Notes in Math. 273, Springer, Berlin, 1972.

[56] B. Sjöberg: On derivatives in topological vector spaces. Nordisk Mat. Tidskr. 14 (1966) 87-96. MR 34 # 4878.

[57] M. Sova: General theory of differentiability in linear topological spaces. Czech. Math. J. 14 (1964) 485-508 (Russian). MR 30 # 2320.

[58] M. Sova: Conditions of differentiability in linear topological spaces. Czech. Math. J. 16 (1966) 339-362 (Russian). MR 33 # 6356.

[59] M.F. Suhinin: Two versions of a theorem on the differentiation of the inverse function in certain linear topological spaces. Vestnik Moskov. Univ. Ser.I, Mat. Meh. 24 (1969) no.6. MR 44 # 5776 a.

[60] M.F. Suhinin: The local invertibility of a differentiable mapping. Uspehi Mat. Nauk 25 (1970) no.5 (155) 249-250. MR 44 # 5776 b.

[61] W. Szczyrba: On differentiability in some class of locally convex spaces. (Short summary). Studia Math. 38 (1970) 458.

[62] W. Szczyrba: Differentiation in locally convex spaces. Studia Math. 39 (1971) 289-306. MR 46 # 6027

[63] M.M. Vainberg and Ya.L. Engel'son: The conditional extremum of functionals in linear topological spaces. Mat. Sb. 45 (1958) 417-422. MR 21 # 1552.

[64] P. Ver Eecke: Sur le calcul différentiel dans les espaces non normés. C.R. Acad. Sci. Paris, Sér.A-B 265 (1967) A720-A723. MR 36 # 6935.

[65] P. Ver Eecke: Quelques observations sur le calcul différentiel. C.R. Acad. Sci. Paris, Sér.A-B 273 (1971) A349-A352. MR 44 # 4519.

[66] M. Wehrli: Differentialrechnung in allgemeinen linearen Räumen I. Rend. Circ. Mat. Palermo, II.Ser. 17 (1968) 81-114.

[67] Y.P. Wong: Differential Calculus and Differentiable Partitions of Unity in Locally Convex Spaces. Univ. of Toronto 1974.

[68] S. Yamamuro: Differential Calculus in Topological Linear Spaces. Lecture Notes in Math. 374, Springer, Berlin, 1974.

[69] S. Yamamuro: Notes on differential calculus in topological linear spaces. To appear in J. Reine Angew. Math.

[70] S. Yamamuro: Notes on differential calculus in topological linear spaces, II. To appear in J. Austral. Math. Soc.

[71] V.I. Averbukh and O.G. Smolyanov: Differentiation and pseudotopologies.(Russian. English summary.) Vestnik Moskov. Univ. Ser. I. Mat. Meh. 27 (1972), no. 1, 3-7. MR 45 # 9130.

[72] V.I. Averbukh and O.G. Smolyanov: Pseudotopologies and differentiation. (Russian. English summary.) Vestnik Moskov. Univ. Ser. I. Mat. Meh. 27 (1972), no. 2, 3-9. MR 45 # 9131.

[73] F. Berquier: Calcul différontiel dans les espaces quasi-bornologiques. Esquisses Math. 20, Paris, 1973.

[74] F. Berquier: Calcul différentiel dans les espaces quasi-bornologiques. C.R. Acad. Sci. Paris, Sér. A, 277 (1973) 897-900.

[75] F. Berquier: Sur la notion d'application différentiable. Cahiers Topologie Géom. Différentielle 14 (1973) 329-338.

[76] W. Gähler: Mengenkonvergenz und verallgemeinerte gleichmässige Konvergenz - ein Beitrag zur Theorie der Differenzierbarkeit. To appear in Math. Nachr.

[77] J.W. Loyd: Differentiable mappings on topological vector spaces. Studia Math. 45 (1973) 147-160.

[78] P. Mankiewicz: On Lipschitz mappings between Fréchet spaces. Studia Math. 41 (1972) 225-241. MR 46 # 7837.

[79] P. Mankiewicz: On the differentiability of Lipschitz mappings in Fréchet spaces. Studia Math. 45 (1973) 15-29.

[80] H. Omori: On the group of diffeomorphisms on a compact manifold. Global Analysis (Proc. Sympos. Pure Math. XV, Berkeley, Calif. 1968) 167-183. AMS, Providence, R.I. 1970. MR 42 # 6864.

[81] H. Omori: Local structures of groups of diffeomorphisms. J. Math. Soc. Japan 24 (1972) 60-88. MR 45 # 4452.

[82] H. Omori: On smooth extension theorems. J. Math. Soc. Japan 24 (1972) 405-432. MR 46 # 4571.

[83] J.P. Penot: Calcul différentiel dans les espaces vectoriels topologiques. Studia Math. 47 (1973) 1-23.

[84] M. Simonnet: Calcul différentiel dans les espaces non normables. Cahiers Topologie Géom. Différentielle 13 (1972) 411-437; 14 (1973) 3-40.

NOTATIONS

v.s.	:	(abbr.) $\mathbb{R}$-vector space
c.v.s.	:	(abbr.) separated convergence vector space (limit vector space)
l.c.s.	:	(abbr.) separated locally convex topological vector space
$\mathbb{N}$	:	set of all non negative integers
$\mathbb{R}$	:	field of all real numbers
$\mathbb{V}$	:	filter of all neighbourhoods of 0 in $\mathbb{R}$
E,F	:	usually: l.c.s.; casually: v.s., c.v.s.
X	:	usually an open set in E; casually a topological space
f	:	usually a function from X into F
Γ_E, Γ_F	:	sets of all continuous semi-norms in E resp. F
α, β	:	elements of Γ_E resp. Γ_F
$\lvert h \rvert_\alpha$	:	value of the semi-norm α at $h \in E$
$\lvert h' \rvert_\beta$	:	value of the semi-norm β at $h' \in F$
$[a,b]$	:	$=\{(1-t)a+tb \mid 0 \leq t \leq 1\}$, segment in a v.s. 1.1.
$L(E,F)$	:	v.s. of all linear maps from E into F 0.0.
$L^n(E,F)$	:	v.s. of all n-linear maps from E^n into F 0.0.
$\Theta^{m,n}$	:	$L^m(E,L^n(E,F)) \to L^{m+n}(E,F)$, the canonical isomorphism of v.s. 0.0.1.
$\mathcal{L}(E,F)$	:	v.s. of all continuous linear maps from E into F 0.0.
$\mathcal{L}^n(E,F)$	:	v.s. of all continuous n-linear maps from E^n into F 0.0.

$\lvert u\rvert_{\beta,S}$	:	$= \sup\{\lvert uh_1\ldots h_n\rvert_\beta \mid h_i \in S,\ 1 \leq i \leq n\}$, where $u \in \mathcal{L}^n(E,F)$, $\beta \in \Gamma_F$, $S \subset E$ 0.1.
$\lvert u\rvert_{\beta,\alpha}$	:	$= \sup\{\lvert uh_1\ldots h_n\rvert_\beta \mid \lvert h_i\rvert_\alpha \leq 1,\ 1 \leq i \leq n\}$, where $u \in \mathcal{L}^n(E,F)$, $\beta \in \Gamma_F$, $\alpha \in \Gamma_E$ 0.1.
Φ	:	$= \Gamma_E^{\Gamma_F}$ 0.6.
$\mathcal{L}^n_\phi(E,F)$	:	$=\{u \in \mathcal{L}^n(E,F) \mid (\forall \beta \in \Gamma_F) \lvert u\rvert_{\beta,\phi(\beta)} < \infty\}$ for $\phi \in \Phi$ 0.6.
$\tilde{\mathcal{L}}^n(E,F)$	:	$=\{u \in \mathcal{L}^n(E,F) \mid (\exists \alpha \in \Gamma_E)(\forall \beta \in \Gamma_F) \lvert u\rvert_{\beta,\alpha} < \infty\}$ 0.7.
ev	:	$\mathcal{L}^n(E,F) \times E^n \to F$, the evaluation map 0.3.
$\tilde{g}$	:	$= ev\circ(g\times id_{E^n}): X \times E^n \to F$, function associated to $g : X \to \mathcal{L}^n(E,F)$ 0.1.2.
Λ	:	convergence structure (limit structure) or topology on a function space, mostly $\mathcal{L}^n(E,F)$ for some $n \in \mathbb{N}$ 0.3.
$\Lambda^\#$	:	equable convergence structure associated to Λ 0.5.
$\mathcal{L}^n_\Lambda(E,F)$	:	$= (\mathcal{L}^n(E,F),\Lambda)$ 0.3.
$\Lambda(u)$	:	set of filters converging to u with respect to Λ 0.3.
$\mathcal{F},\mathcal{G}$	:	filters on a function space
$\mathcal{B},\mathcal{R},\mathcal{X}$	:	filters on E
$\mathcal{U}$	:	filter of all neighbourhoods of 0 in E
$[h],[S]$	:	filters on E generated by $\{h\}$ and S resp. for $h \in E$, $S \subset E$
$\mathfrak{S}$	:	covering of E consisting of bounded sets 0.1.
$\mathfrak{S}_s$	:	collection of all finite subsets of E
$\mathfrak{S}_k$	:	collection of all compact subsets of E
$\mathfrak{S}_{pk}$	:	collection of all precompact subsets of E
$\mathfrak{S}_b$	:	collection of all bounded subsets of E
$\Lambda_\mathfrak{S}$	:	topology of $\mathfrak{S}$-convergence 0.1.

Λ_s : $= \Lambda_{\mathfrak{S}_s}$, topology of simple convergence 0.1.

Λ_k : $= \Lambda_{\mathfrak{S}_k}$, topology of compact convergence 0.1.

Λ_{pk} : $= \Lambda_{\mathfrak{S}_{pk}}$, topology of precompact convergence 0.1.

Λ_b : $= \Lambda_{\mathfrak{S}_b}$, topology of bounded convergence 0.1.

$\mathcal{L}^n_{\mathfrak{S}}(E,F)$: $=(\mathcal{L}^n(E,F),\Lambda_{\mathfrak{S}})$ 0.1.

$L^n_s(E,F)$: $=(L^n(E,F),\Lambda_s)$ 0.1.

$\mathcal{L}^n_s(E,F)$: $=(\mathcal{L}^n(E,F),\Lambda_s)$ 0.1.

$\mathcal{L}^n_k(E,F)$: $=(\mathcal{L}^n(E,F),\Lambda_k)$ 0.1.

$\mathcal{L}^n_{pk}(E,F)$: $=(\mathcal{L}^n(E,F),\Lambda_{pk})$ 0.1.

$\mathcal{L}^n_b(E,F)$: $=(\mathcal{L}^n(E,F),\Lambda_b)$ 0.1.

$\mathcal{H}^n_{\mathfrak{S}}(E,F)$: $= \mathcal{L}_{\mathfrak{S}}(E,\mathcal{H}^{n-1}_{\mathfrak{S}}(E,F))$ if $n \geq 1$, $\mathcal{H}^0_{\mathfrak{S}}(E,F) := F$ 0.2.

$\mathcal{H}^n_s(E,F)$: $= \mathcal{H}^n_{\mathfrak{S}_s}(E,F)$ 0.2.

$\mathcal{H}^n_k(E,F)$: $= \mathcal{H}^n_{\mathfrak{S}_k}(E,F)$ 0.2.

$\mathcal{H}^n_{pk}(E,F)$: $= \mathcal{H}^n_{\mathfrak{S}_{pk}}(E,F)$ 0.2.

$\mathcal{H}^n_b(E,F)$: $= \mathcal{H}^n_{\mathfrak{S}_b}(E,F)$ 0.2.

Λ_c : structure of continuous convergence on $\mathcal{L}^n(E,F)$ 0.3.

Λ_{qb} : structure of quasi-bounded convergence on $\mathcal{L}^n(E,F)$ 0.4.

Π,Θ : special convergence structures on $\mathcal{L}^n(E,F)$ 0.5.,0.7.

Δ : Marinescu's convergence structure on $\mathcal{L}^n(E,F)$ 0.6.

$\mathcal{L}^n_c(E,F)$: $= (\mathcal{L}^n(E,F),\Lambda_c)$ 0.3.

$\mathcal{L}^n_{qb}(E,F)$: $= (\mathcal{L}^n(E,F),\Lambda_{qb})$ 0.4.

$\mathcal{L}^n_{\Pi}(E,F)$: $= (\mathcal{L}^n(E,F),\Pi)$ 0.5.

$\mathcal{L}^n_{\Delta}(E,F)$: $= (\mathcal{L}^n(E,F),\Delta)$ 0.6.

$\mathcal{L}^n_{\Theta}(E,F)$: $= (\mathcal{L}^n(E,F),\Theta)$ 0.7.

C^p_Λ : differentiability class of order p with respect to the convergence structure Λ 1.0.0., 2.6.0.

$C^p_{\mathfrak{S}}$: differentiability class of order p with respect to the covering $\mathfrak{S}$ 2.5.0.

$C^0(X,F)$: v.s. of all continuous functions from X into F 1.0.

$C^p(X,F)$: v.s. of all functions of class C^p from X into F, X open in $\mathbb{R}^m$ 1.0.6., 2.3.1.

$C^p_\Lambda(X,F)$: v.s. of all functions of class C^p_Λ from X into F 1.0.0., 2.6.0.

$C^p_{\mathfrak{S}}(X,F)$: v.s. of all functions of class $C^p_{\mathfrak{S}}$ from X into F 2.5.0.

$C^p_s(X,F)$: $= C^p_{\mathfrak{S}_s}(X,F) = C^p_{\Lambda_s}(X,F)$ 1.0., 2.5.

$C^p_k(X,F)$: $= C^p_{\mathfrak{S}_k}(X,F) = C^p_{\Lambda_k}(X,F)$ 1.0., 2.5.

$C^p_{pk}(X,F)$: $= C^p_{\mathfrak{S}_{pk}}(X,F) = C^p_{\Lambda_{pk}}(X,F)$ 1.0., 2.5.

$C^p_b(X,F)$: $= C^p_{\mathfrak{S}_b}(X,F) = C^p_{\Lambda_b}(X,F)$ 1.0., 2.5.

$C^p_c(X,F)$: $= C^p_{\Lambda_c}(X,F)$ 1.0., 2.6.

$C^p_{qb}(X,F)$: $= C^p_{\Lambda_{qb}}(X,F)$ 1.0., 2.6.

$C^\infty_\Lambda(X,F)$: $= \bigcap_{p\in\mathbb{N}} C^p_\Lambda(X,F)$ 2.9.0.

$C^\infty(X,F)$: $= C^\infty_s(X,F) = C^\infty_\Theta(X,F) = C^\infty_\Delta(X,F)$ X open in E, if E a Banach space or if E a Fréchet space and F normable 2.9.5.

J : open set in $\mathbb{R}$

$f'(t)$: $= \lim_{\tau\to 0} \tau^{-1}(f(t+\tau)-f(t))$, derivative of $f : J \to F$, $J \subset \mathbb{R}$, at $t \in J$ A.1.1.

$f^{(r)}$	:	$= (f^{(r-1)})'$ if $r \geq 1$, $f^{(0)} = f$ if $f : J \to F$ is of class C^r A.1.2.
$\int_a^b f$	:	$= \int_a^b f(t)dt$ Riemann Integral of a (continuous) function $f : [a,b] \to F$, $[a,b] \subset \mathbb{R}$. A.2.1.
Df	:	derivative of $f : X \to F$ 1.0.0.
$D^p f$	:	derivative of order p of $f : X \to F$ 2.1.0.
$Rf_x(h)$	:	$= Rf(x,h) := f(x+h) - f(x) - Df(x)h$, remainder of f at x in the direction h 1.1.0.
$R_p f_x(h)$	:	$= R_p f(x,h)$, remainder of order p of f at x in the direction h 2.2.0.
$\Theta Rf_x(t,h)$	:	$= t^{-1} Rf_x(th)$ if $t \neq 0$; $\Theta Rf_x(0,h) := 0$ 1.1.
$\Theta_p R_p f_x(t,h)$	:	$= t^{-p} R_p f_x(th)$ if $t \neq 0$; $\Theta_p R_p f_x(0,h) := 0$ 2.8.0.
$J_{x,h}$	:	$= \{t \in \mathbb{R} \mid x+th \in X\}$, $X \subset E$, $(x,h) \in E \times E$ 1.1.1.
Ω_X	:	$= \{(x,h) \in X \times E \mid x+h \in X\}$, $X \subset E$ 1.1.
Ω_x	:	$= \{(t,h) \in \mathbb{R} \times E \mid x+th \in X\}$, $X \subset E$, $x \in E$ 1.1.

Vol. 247: Lectures on Operator Algebras. Tulane University Ring and Operator Theory Year, 1970–1971. Volume II. XI, 786 pages. 1972. DM 40,–

Vol. 248: Lectures on the Applications of Sheaves to Ring Theory. Tulane University Ring and Operator Theory Year, 1970–1971. Volume III. VIII, 315 pages. 1971. DM 26,–

Vol. 249: Symposium on Algebraic Topology. Edited by P. J. Hilton. VII, 111 pages. 1971. DM 16,–

Vol. 250: B. Jónsson, Topics in Universal Algebra. VI, 220 pages. 1972. DM 20,–

Vol. 251: The Theory of Arithmetic Functions. Edited by A. A. Gioia and D. L. Goldsmith VI, 287 pages. 1972. DM 24,–

Vol. 252: D. A. Stone, Stratified Polyhedra. IX, 193 pages. 1972. DM 18,–

Vol. 253: V. Komkov, Optimal Control Theory for the Damping of Vibrations of Simple Elastic Systems. V, 240 pages. 1972. DM 20,–

Vol. 254: C. U. Jensen, Les Foncteurs Dérivés de $\varprojlim$ et leurs Applications en Théorie des Modules. V, 103 pages. 1972. DM 16,–

Vol. 255: Conference in Mathematical Logic – London '70. Edited by W. Hodges. VIII, 351 pages. 1972. DM 26,–

Vol. 256: C. A. Berenstein and M. A. Dostal, Analytically Uniform Spaces and their Applications to Convolution Equations. VII, 130 pages. 1972. DM 16,–

Vol. 257: R. B. Holmes, A Course on Optimization and Best Approximation. VIII, 233 pages. 1972. DM 20,–

Vol. 258: Séminaire de Probabilités VI. Edited by P. A. Meyer. VI, 253 pages. 1972. DM 22,–

Vol. 259: N. Moulis, Structures de Fredholm sur les Variétés Hilbertiennes. V, 123 pages. 1972. DM 16,–

Vol. 260: R. Godement and H. Jacquet, Zeta Functions of Simple Algebras. IX, 188 pages. 1972. DM 18,–

Vol. 261: A. Guichardet, Symmetric Hilbert Spaces and Related Topics. V, 197 pages. 1972. DM 18,–

Vol. 262: H. G. Zimmer, Computational Problems, Methods, and Results in Algebraic Number Theory. V, 103 pages. 1972. DM 16,–

Vol. 263: T. Parthasarathy, Selection Theorems and their Applications. VII, 101 pages. 1972. DM 16,–

Vol. 264: W. Messing, The Crystals Associated to Barsotti-Tate Groups: With Applications to Abelian Schemes. III, 190 pages. 1972. DM 18,–

Vol. 265: N. Saavedra Rivano, Catégories Tannakiennes. II, 418 pages. 1972. DM 26,–

Vol. 266: Conference on Harmonic Analysis. Edited by D. Gulick and R. L. Lipsman. VI, 323 pages. 1972. DM 24,–

Vol. 267: Numerische Lösung nichtlinearer partieller Differential- und Integro-Differentialgleichungen. Herausgegeben von R. Ansorge und W. Törnig. VI, 339 Seiten. 1972. DM 26,–

Vol. 268: C. G. Simader, On Dirichlet's Boundary Value Problem. IV, 238 pages. 1972. DM 20,–

Vol. 269: Théorie des Topos et Cohomologie Etale des Schémas. (SGA 4). Dirigé par M. Artin, A. Grothendieck et J. L. Verdier. XIX, 525 pages. 1972. DM 50,–

Vol. 270: Théorie des Topos et Cohomologie Etale des Schémas. Tome 2. (SGA 4). Dirigé par M. Artin, A. Grothendieck et J. L. Verdier. V, 418 pages. 1972. DM 50,–

Vol. 271: J. P. May, The Geometry of Iterated Loop Spaces. IX, 175 pages. 1972. DM 18,–

Vol. 272: K. R. Parthasarathy and K. Schmidt, Positive Definite Kernels, Continuous Tensor Products, and Central Limit Theorems of Probability Theory. VI, 107 pages. 1972. DM 16,–

Vol. 273: U. Seip, Kompakt erzeugte Vektorräume und Analysis. IX, 119 Seiten. 1972. DM 16,–

Vol. 274: Toposes, Algebraic Geometry and Logic. Edited by. F. W. Lawvere. VI, 189 pages. 1972. DM 18,–

Vol. 275: Séminaire Pierre Lelong (Analyse) Année 1970–1971. VI, 181 pages. 1972. DM 18,–

Vol. 276: A. Borel, Représentations de Groupes Localement Compacts. V, 98 pages. 1972. DM 16,–

Vol. 277: Séminaire Banach. Edité par C. Houzel. VII, 229 pages. 1972. DM 20,–

Vol. 278: H. Jacquet, Automorphic Forms on GL(2). Part II. XIII, 142 pages. 1972. DM 16,–

Vol. 279: R. Bott, S. Gitler and I. M. James, Lectures on Algebraic and Differential Topology. V, 174 pages. 1972. DM 18,–

Vol. 280: Conference on the Theory of Ordinary and Partial Differential Equations. Edited by W. N. Everitt and B. D. Sleeman. XV, 367 pages. 1972. DM 26,–

Vol. 281: Coherence in Categories. Edited by S. Mac Lane. VII, 235 pages. 1972. DM 20,–

Vol. 282: W. Klingenberg und P. Flaschel, Riemannsche Hilbertmannigfaltigkeiten. Periodische Geodätische. VII, 211 Seiten. 1972. DM 20,–

Vol. 283: L. Illusie, Complexe Cotangent et Déformations II. VII, 304 pages. 1972. DM 24,–

Vol. 284: P. A. Meyer, Martingales and Stochastic Integrals I. VI, 89 pages. 1972. DM 16,–

Vol. 285: P. de la Harpe, Classical Banach-Lie Algebras and Banach-Lie Groups of Operators in Hilbert Space. III, 160 pages. 1972. DM 16,–

Vol. 286: S. Murakami, On Automorphisms of Siegel Domains. V, 95 pages. 1972. DM 16,–

Vol. 287: Hyperfunctions and Pseudo-Differential Equations. Edited by H. Komatsu. VII, 529 pages. 1973. DM 36,–

Vol. 288: Groupes de Monodromie en Géométrie Algébrique. (SGA 7 I). Dirigé par A. Grothendieck. IX, 523 pages. 1972. DM 50,–

Vol. 289: B. Fuglede, Finely Harmonic Functions. III, 188. 1972. DM 18,–

Vol. 290: D. B. Zagier, Equivariant Pontrjagin Classes and Applications to Orbit Spaces. IX, 130 pages. 1972. DM 16,–

Vol. 291: P. Orlik, Seifert Manifolds. VIII, 155 pages. 1972. DM 16,–

Vol. 292: W. D. Wallis, A. P. Street and J. S. Wallis, Combinatorics: Room Squares, Sum-Free Sets, Hadamard Matrices. V, 508 pages. 1972. DM 50,–

Vol. 293: R. A. DeVore, The Approximation of Continuous Functions by Positive Linear Operators. VIII, 289 pages. 1972. DM 24,–

Vol. 294: Stability of Stochastic Dynamical Systems. Edited by R. F. Curtain. IX, 332 pages. 1972. DM 26,–

Vol. 295: C. Dellacherie, Ensembles Analytiques, Capacités, Mesures de Hausdorff. XII, 123 pages. 1972. DM 16,–

Vol. 296: Probability and Information Theory II. Edited by M. Behara, K. Krickeberg and J. Wolfowitz. V, 223 pages. 1973. DM 20,–

Vol. 297: J. Garnett, Analytic Capacity and Measure. IV, 138 pages. 1972. DM 16,–

Vol. 298: Proceedings of the Second Conference on Compact Transformation Groups. Part 1. XIII, 453 pages. 1972. DM 32,–

Vol. 299: Proceedings of the Second Conference on Compact Transformation Groups. Part 2. XIV, 327 pages. 1972. DM 26,–

Vol. 300: P. Eymard, Moyennes Invariantes et Représentations Unitaires. II. 113 pages. 1972. DM 16,–

Vol. 301: F. Pittnauer, Vorlesungen über asymptotische Reihen. VI, 186 Seiten. 1972. DM 18,–

Vol. 302: M. Demazure, Lectures on p-Divisible Groups. V, 98 pages. 1972. DM 16,–

Vol. 303: Graph Theory and Applications. Edited by Y. Alavi, D. R. Lick and A. T. White. IX, 329 pages. 1972. DM 26,–

Vol. 304: A. K. Bousfield and D. M. Kan, Homotopy Limits, Completions and Localizations. V, 348 pages. 1972. DM 26,–

Vol. 305: Théorie des Topos et Cohomologie Etale des Schémas. Tome 3. (SGA 4). Dirigé par M. Artin, A. Grothendieck et J. L. Verdier. VI, 640 pages. 1973. DM 50,–

Vol. 306: H. Luckhardt, Extensional Gödel Functional Interpretation. VI, 161 pages. 1973. DM 18,–

Vol. 307: J. L. Bretagnolle, S. D. Chatterji et P.-A. Meyer, Ecole d'été de Probabilités: Processus Stochastiques. VI, 198 pages. 1973. DM 20,–

Vol. 308: D. Knutson, λ-Rings and the Representation Theory of the Symmetric Group. IV, 203 pages. 1973. DM 20,–

Vol. 309: D. H. Sattinger, Topics in Stability and Bifurcation Theory. VI, 190 pages. 1973. DM 18,–

Vol. 310: B. Iversen, Generic Local Structure of the Morphisms in Commutative Algebra. IV, 108 pages. 1973. DM 16,-

Vol. 311: Conference on Commutative Algebra. Edited by J. W. Brewer and E. A. Rutter. VII, 251 pages. 1973. DM 22,-

Vol. 312: Symposium on Ordinary Differential Equations. Edited by W. A. Harris, Jr. and Y. Sibuya. VIII, 204 pages. 1973. DM 22,-

Vol. 313: K. Jörgens and J. Weidmann, Spectral Properties of Hamiltonian Operators. III, 140 pages. 1973. DM 16,-

Vol. 314: M. Deuring, Lectures on the Theory of Algebraic Functions of One Variable. VI, 151 pages. 1973. DM 16,-

Vol. 315: K. Bichteler, Integration Theory (with Special Attention to Vector Measures). VI, 357 pages. 1973. DM 26,-

Vol. 316: Symposium on Non-Well-Posed Problems and Logarithmic Convexity. Edited by R. J. Knops. V, 176 pages. 1973. DM 18,-

Vol. 317: Séminaire Bourbaki – vol. 1971/72. Exposés 400–417. IV, 361 pages. 1973. DM 26,-

Vol. 318: Recent Advances in Topological Dynamics. Edited by A. Beck, VIII, 285 pages. 1973. DM 24,-

Vol. 319: Conference on Group Theory. Edited by R. W. Gatterdam and K. W. Weston. V, 188 pages. 1973. DM 18,-

Vol. 320: Modular Functions of One Variable I. Edited by W. Kuyk. V, 195 pages. 1973. DM 18,-

Vol. 321: Séminaire de Probabilités VII. Edité par P. A. Meyer. VI, 322 pages. 1973. DM 26,-

Vol. 322: Nonlinear Problems in the Physical Sciences and Biology. Edited by I. Stakgold, D. D. Joseph and D. H. Sattinger. VIII, 357 pages. 1973. DM 26,-

Vol. 323: J. L. Lions, Perturbations Singulières dans les Problèmes aux Limites et en Contrôle Optimal. XII, 645 pages. 1973. DM 42,-

Vol. 324: K. Kreith, Oscillation Theory. VI, 109 pages. 1973. DM 16,-

Vol. 325: Ch.-Ch. Chou, La Transformation de Fourier Complexe et L'Equation de Convolution. IX, 137 pages. 1973. DM 16,-

Vol. 326: A. Robert, Elliptic Curves. VIII, 264 pages. 1973. DM 22,-

Vol. 327: E. Matlis, 1-Dimensional Cohen-Macaulay Rings. XII, 157 pages. 1973. DM 18,-

Vol. 328: J. R. Büchi and D. Siefkes, The Monadic Second Order Theory of All Countable Ordinals. VI, 217 pages. 1973. DM 20,-

Vol. 329: W. Trebels, Multipliers for (C, α)-Bounded Fourier Expansions in Banach Spaces and Approximation Theory. VII, 103 pages. 1973. DM 16,-

Vol. 330: Proceedings of the Second Japan-USSR Symposium on Probability Theory. Edited by G. Maruyama and Yu. V. Prokhorov. VI, 550 pages. 1973. DM 36,-

Vol. 331: Summer School on Topological Vector Spaces. Edited by L. Waelbroeck. VI, 226 pages. 1973. DM 20,-

Vol. 332: Séminaire Pierre Lelong (Analyse) Année 1971-1972. V, 131 pages. 1973. DM 16,-

Vol. 333: Numerische, insbesondere approximationstheoretische Behandlung von Funktionalgleichungen. Herausgegeben von R. Ansorge und W. Törnig. VI, 296 Seiten. 1973. DM 24,-

Vol. 334: F. Schweiger, The Metrical Theory of Jacobi-Perron Algorithm. V, 111 pages. 1973. DM 16,-

Vol. 335: H. Huck, R. Roitzsch, U. Simon, W. Vortisch, R. Walden, B. Wegner und W. Wendland, Beweismethoden der Differentialgeometrie im Großen. IX, 159 Seiten. 1973. DM 18,-

Vol. 336: L'Analyse Harmonique dans le Domaine Complexe. Edité par E. J. Akutowicz. VIII, 169 pages. 1973. DM 18,-

Vol. 337: Cambridge Summer School in Mathematical Logic. Edited by A. R. D. Mathias and H. Rogers. IX, 660 pages. 1973. DM 42,-

Vol: 338: J. Lindenstrauss and L. Tzafriri, Classical Banach Spaces. IX, 243 pages. 1973. DM 22,-

Vol. 339: G. Kempf, F. Knudsen, D. Mumford and B. Saint-Donat, Toroidal Embeddings I. VIII, 209 pages. 1973. DM 20,-

Vol. 340: Groupes de Monodromie en Géométrie Algébrique. (SGA 7 II). Par P. Deligne et N. Katz. X, 438 pages. 1973. DM 40,-

Vol. 341: Algebraic K-Theory I, Higher K-Theories. Edited by H. Bass. XV, 335 pages. 1973. DM 26,-

Vol. 342: Algebraic K-Theory II, "Classical" Algebraic K-Theory, and Connections with Arithmetic. Edited by H. Bass. XV, 527 pages. 1973. DM 36,-

Vol. 343: Algebraic K-Theory III, Hermitian K-Theory and Geometric Applications. Edited by H. Bass. XV, 572 pages. 1973. DM 38,-

Vol. 344: A. S. Troelstra (Editor), Metamathematical Investigation of Intuitionistic Arithmetic and Analysis. XVII, 485 pages. 1973. DM 34,-

Vol. 345: Proceedings of a Conference on Operator Theory. Edited by P. A. Fillmore. VI, 228 pages. 1973. DM 20,-

Vol. 346: Fučík et al., Spectral Analysis of Nonlinear Operators. II, 287 pages. 1973. DM 26,-

Vol. 347: J. M. Boardman and R. M. Vogt, Homotopy Invariant Algebraic Structures on Topological Spaces. X, 257 pages. 1973. DM 22,-

Vol. 348: A. M. Mathai and R. K. Saxena, Generalized Hypergeometric Functions with Applications in Statistics and Physical Sciences. VII, 314 pages. 1973. DM 26,-

Vol. 349: Modular Functions of One Variable II. Edited by W. Kuyk and P. Deligne. V, 598 pages. 1973. DM 38,-

Vol. 350: Modular Functions of One Variable III. Edited by W. Kuyk and J.-P. Serre. V, 350 pages. 1973. DM 26,-

Vol. 351: H. Tachikawa, Quasi-Frobenius Rings and Generalizations. XI, 172 pages. 1973. DM 18,-

Vol. 352: J. D. Fay, Theta Functions on Riemann Surfaces. V, 137 pages. 1973. DM 16,-

Vol. 353: Proceedings of the Conference on Orders, Group Rings and Related Topics. Organized by J. S. Hsia, M. L. Madan and T. G. Ralley. X, 224 pages. 1973. DM 20,-

Vol. 354: K. J. Devlin, Aspects of Constructibility. XII, 240 pages. 1973. DM 22,-

Vol. 355: M. Sion, A Theory of Semigroup Valued Measures. V, 140 pages. 1973. DM 16,-

Vol. 356: W. L. J. van der Kallen, Infinitesimally Central-Extensions of Chevalley Groups. VII, 147 pages. 1973. DM 16,-

Vol. 357: W. Borho, P. Gabriel und R. Rentschler, Primideale in Einhüllenden auflösbarer Lie-Algebren. V, 182 Seiten. 1973. DM 18,-

Vol. 358: F. L. Williams, Tensor Products of Principal Series Representations. VI, 132 pages. 1973. DM 16,-

Vol. 359: U. Stammbach, Homology in Group Theory. VIII, 183 pages. 1973. DM 18,-

Vol. 360: W. J. Padgett and R. L. Taylor, Laws of Large Numbers for Normed Linear Spaces and Certain Fréchet Spaces. VI, 111 pages. 1973. DM 16,-

Vol. 361: J. W. Schutz, Foundations of Special Relativity: Kinematic Axioms for Minkowski Space Time. XX, 314 pages. 1973. DM 26,-

Vol. 362: Proceedings of the Conference on Numerical Solution of Ordinary Differential Equations. Edited by D. Bettis. VIII, 490 pages. 1974. DM 34,-

Vol. 363: Conference on the Numerical Solution of Differential Equations. Edited by G. A. Watson. IX, 221 pages. 1974. DM 20,-

Vol. 364: Proceedings on Infinite Dimensional Holomorphy. Edited by T. L. Hayden and T. J. Suffridge. VII, 212 pages. 1974. DM 20,-

Vol. 365: R. P. Gilbert, Constructive Methods for Elliptic Equations. VII, 397 pages. 1974. DM 26,-

Vol. 366: R. Steinberg, Conjugacy Classes in Algebraic Groups (Notes by V. V. Deodhar). VI, 159 pages. 1974. DM 18,-

Vol. 367: K. Langmann und W. Lütkebohmert, Cousinverteilungen und Fortsetzungssätze. VI, 151 Seiten. 1974. DM 16,-

Vol. 368: R. J. Milgram, Unstable Homotopy from the Stable Point of View. V, 109 pages. 1974. DM 16,-

Vol. 369: Victoria Symposium on Nonstandard Analysis. Edited by A. Hurd and P. Loeb. XVIII, 339 pages. 1974. DM 26,-

Vol. 370: B. Mazur and W. Messing, Universal Extensions and One Dimensional Crystalline Cohomology. VII, 134 pages. 1974. DM 16,-